Francis A. Drexel
LIBRARY

Books for College Libraries

Third Edition

Core Collection

SAINT JOSEPH'S UNIVERSITY

Bargaining for Job Safety and Health

Bargaining for Job Safety and Health

Lawrence S. Bacow

HD
7262.5
$.U6$
$B32$

The MIT Press
Cambridge, Massachusetts,
and London, England

Second printing, February 1981
© 1980 by
The Massachusetts Institute of Technology

All rights reserved. No part of this book may be reproduced in any form or by any means, electronic or mechanical, including photocopying, recording, or by any information storage and retrieval system, without permission in writing from the publisher.

This book was set in Fototronic Zenith
by The Colonial Cooperative Press Inc.
and printed and bound by Halliday Lithograph
in the United States of America.

Library of Congress Cataloging in Publication Data

Bacow, Lawrence S.
   Bargaining for job safety and health.

   Bibliography: p. 150
   Includes index.
   1. Industrial safety—United States.
2. Industrial hygiene—United States.
3. Collective bargaining—United States.   I. Title.
HD7262.5.U6B32   331.89   80-16602
ISBN 0-262-02152-8

To Adele

Contents

Preface

I
THE PROBLEMS OF REGULATION

1
Scope and Complexity   3

2
"Command and Control" Strategies   12

II
EVALUATING POLICY

3
The Impact of OSHA   24

4
Doing Better   51

III
BARGAINING AS A REGULATORY STRATEGY

5
From Theory to Practice   60

6
Influences on the Likelihood of Health and Safety Bargaining   88

7
Ways to Encourage Labor-Management Health and Safety Activity   103

8
Lessons for the Design of Regulatory Policy   122

Notes   133

Selective Bibliography   150

Index   155

Preface

As the title suggests, this book is about how labor and management bargain over hazards in the workplace. This subject has received relatively little attention to date in the growing literature on occupational safety and health policy, because we tend to think separately about government and private-sector activity to control job hazards. My premise is that public policy toward job safety and health can be improved if we begin to think about public and private-sector activity as complementary. This book analyzes how government might capitalize on the potential capacity of labor and management to abate job hazards on their own through collective bargaining processes.

Occupational safety and health is a particularly rich topic for scholars interested in the design of regulatory policy. Virtually all of the major issues that constitute the debate over regulatory reform become evident in discussions about policy toward job hazards. As a result, many of the insights gained in the analysis of this regulatory problem have application to others. I have tried to highlight generic issues by including sections that discuss how these issues arise in other contexts. Readers interested in other regulatory problems will want to pay close attention to chapters 1, 2, and 8.

The book is divided into three parts. Part I introduces readers to the technical, economic, political, and administrative problems that must be overcome if job hazards are to be controlled effectively. Part II evaluates the performance of the Occupational Safety and Health Administration in addressing these problems, and reviews suggestions for reform. Part III explores one suggested approach—collective bargaining—in depth, through a series of case studies, and offers some ideas on how this approach might be pursued in practice.

The idea for this book came from Henry Perritt, former Deputy Undersecretary of Labor for Economic Policy Review, and Roland Droitch, Associate Deputy Undersecretary. In addition to suggesting the topic, they also helped to establish contacts within the Labor Department and made many useful suggestions about the direction and scope of the analysis.

A number of colleagues at MIT and Harvard commented on different sections of this manuscript. I am indebted to Leslie Boden, Joseph Ferreira, Bernard Frieden, Valerie Nelson, Albert Nichols, and Richard Zeckhauser for their suggestions. Steven Kelman graciously shared with me some of the preliminary results of his survey of workers.

Special thanks are owed to Michael O'Hare and Mark Moore, who greatly influenced the way I think about regulatory problems through an extended series of informal discussions. Both also commented extensively on an early draft.

Many people in labor, management, and government gave freely of their time to be interviewed. They are too numerous to mention. I am grateful to all of them.

Financial support for this research was provided in part by the U.S. Department of Labor and the Harvard Faculty Project on Regulation. The Department of Urban Studies and Planning at MIT provided needed secretarial support. Penny Johnson and Jeanne Winbush assisted in the preparation of the manuscript.

Finally, I would like to thank my wife Adele and my son Jay. The prospect of spending my time with them gave me a terrific incentive to finish this book quickly.

1
THE PROBLEMS OF REGULATION

1
Scope and Complexity

In late December 1977 a grain elevator exploded in Westwego, Louisiana. The explosion was caused by a spark that ignited an excessive accumulation of grain dust. Thirty-five workers died in the ensuing fire. In April 1978 the scaffolding supporting workers constructing a cooling tower for a Virginia nuclear plant collapsed. The anchor bolts supporting the scaffold failed because they had been set into inadequately cured concrete. Fifty-one workers died. Recently a Senate subcommittee held hearings on the incidence of lung cancer among uranium miners. It appears that these workers die from lung cancer at nearly four times the national average rate for men of the same age.

Few would deny that the United States has an occupational safety and health problem. Although good statistics are hard to come by, the National Safety Council estimates that each year roughly 13,000 Americans die in job-related accidents. It is likely that these figures grossly underestimate the true number of job-related fatalities. Because occupational diseases have long latent periods and are difficult to diagnose, many job-related fatalities are never reported. Estimates of these unreported deaths range up to 100,000 per year.[1] If these estimates are accurate, the annual number of occupational deaths is twice that of traffic fatalities, four times that of deaths resulting from accidents in the home, and five times that of homicides.

The cost of job-related morbidity and mortality is staggering. Occasionally, family members suffer not only grief but disease as well. Workers may carry carcinogenic dust home on their clothes. Some chemicals threaten the fragile existence of the unborn. Even if we could eliminate the pain and suffering associated with occupational disability, job hazards would still constitute a major social problem. Job-related accidents and illnesses impair productivity and tax the already overburdened health-care system. The National Safety Council estimates the non–pain-and-suffering costs of occupational disability at nearly one percent of the gross national product, or over $16 billion annually.[2]

It is difficult to ascertain whether the occupational safety and health situation is getting better or worse. The statistics tell a mixed story. From 1957 to 1970, in manufacturing the lost-time injury rate per million hours worked is said to have increased from 11.1 to

4
Scope and Complexity

15.2.[3] However, recent analysis casts doubt on the accuracy of these data. Injury rates are sensitive to changes in worker fatigue, worker experience, and pace of production. Adjustment for seasonal changes in overtime, hiring rates, and capacity utilization makes it even more difficult to determine whether the injury rate is going up or down. Depending upon when the trend analysis is started, it is possible to conclude either that there is no trend after cyclical influences are accounted for or that there is a sharp upward trend.[4]

We also cannot determine whether the occupational safety and health situation is getting better or worse merely by looking at the effect of changes in production technology. On one hand, automation and improvements in industrial hygiene have removed many workers from positions of risk. On the other hand, technological advances account for the synthesis of over 3,000 new chemicals per year, each potentially harmful. About all that we can safely conclude from looking at statistics and changes in technology is that occupational safety and health is a big problem that may be getting bigger.

Regulating job hazards is even harder than measuring them. Four characteristics of the occupational safety and health problem make regulation especially vexing.

First, our ignorance about occupational risks is profound. We are not sure how many workers are killed or disabled by job hazards. We do not know what causes many accidents on the job. We are not certain which chemicals are hazardous, or what the effects of mixing otherwise safe substances may be. We do not know the dose-response relationship for known hazardous substances. We often do not know how to measure the benefits from hazard abatement. And we are unable to predict with accuracy the costs of different strategies for eliminating hazards. It is difficult to allocate resources effectively to abate hazards without resolving at least some of these uncertainties.

Second, we lack a clear consensus on the appropriate normative criteria to be used in setting policy. We cannot agree on how to value the benefits from hazard abatement. We cannot agree on what role, if any, considerations of cost should play in determining how to make the workplace safe. We cannot agree on what consti-

5
Scope and Complexity

tutes an acceptable risk. And we often cannot agree on how to allocate responsibility among labor, management, and government for making decisions about steps to be taken in eliminating hazards from the workplace.

Third, making employment safe in this country involves changing the behavior of an extraordinarily large number of people and institutions. Sixty million workers are employed in the United States in more than five million workplaces distributed over a very large area. To make each workplace safe requires physical changes in each work environment as well as changes in the day-to-day habits of workers and their supervisors.

Finally, workplaces differ—sometimes only slightly and sometimes drastically. Some workplaces are large; some are small. Some produce cars; others produce deposits and withdrawals. Some are inherently dangerous; some inherently safe. Some workers worry about back sprains; others about cancer. Some firms are capital-intensive; others labor-intensive. Some are unionized; others are not. In some industries (such as steel) a worker may be employed by the same company in the same plant for 30 years; in others (such as construction) a worker may work at fifty different sites each year. Each firm has a different production function, cost function, production technology, market position, and ability (and willingness) to respond to government regulation. In effect, each firm has a slightly different health and safety problem. The challenge of regulation is to design a regulatory policy that responds effectively to the diverse conditions encountered among diverse employment situations.

To fully appreciate the complexity of trying to regulate occupational safety and health, let us compare it against the regulatory dimensions of another major public health and safety problem: polio. Before the discovery of an effective vaccine, polio struck without warning, mainly among the young, and left many of its victims paralyzed for life. In 1952 there were over 58,000 reported cases of acute poliomyelitis.[5] Discovery of the Salk vaccine was announced in April 1955. Only 5,787 cases of polio were reported in 1957, and by the mid-1960s fewer than 100 cases were being reported annually. Why was it relatively easy to get rid of the polio problem? Because each person faced essentially the same type of risk, and therefore the problem was amenable to control by a single solution: the vaccine. In contrast, to solve the occupational safety and health problem we must find a way to eliminate hundreds of thousands of different risks. Similarly, from an immunological perspective the potential polio population was nearly perfectly homogeneous; the same vaccine was effective for almost every person. But workplaces are extremely diverse in ways that influence the effectiveness of hazard-abatement strategies; what may be effective in one may fail in another. To be truly effective, health and safety programs need to be specially tailored for each workplace.

If regulated institutions were perfectly homogeneous, the design of regulatory policy would be a relatively simple task: The regulator would merely study a sample regulatee, determine the actions needed to produce the desired outcome, and then mandate those actions nationwide. Then, the regulator would only have to observe the actual effectiveness of the regulatory program in one firm to monitor its effectiveness throughout the country. Moreover, by observing just one firm the regulator would learn how to adapt the policy to changes in external conditions that affected the success of the overall regulatory program. Enforcement would also be simple. What worked in one firm would work in every firm.

We do not live in the "dream world" just described. Some examples may help to illustrate how the diversity of problems, regulatees, and environments complicates the design of regulatory policy for problems other than job safety and health.

Diversity of Hazards

Consider the problem of regulating consumer-product safety. Millions of products are marketed annually in the United States by thousands of different producers. The risk associated with each product varies with its design; the quality the materials and the workmanship; the directions provided to the user; the user's skill, judgment, and caution; the age of the product; whether the product is being used for its intended purpose; the way the product has been maintained; and the extent to which the product is used with other products. If we wanted to evaluate alternatives for making just one product safe, we would want to gather information about

7
Scope and Complexity

the size of the population at risk, the cost of different product designs, and the likely reduction in risk for each design. Rarely does a regulator have complete access to this information; much of it must be obtained from the producer. Regulating product safety on a product-by-product basis places extraordinary informational and analytic demands on the regulator.

On close inspection, even problems that appear to be all of a piece reveal themselves to be multifaceted. For example, reclaiming land from which coal has been strip-mined is a fundamentally different operation than reclaiming land from which copper has been taken. Coal is usually found in dense veins just below the topsoil. When it is removed, there usually is very little material left (other than the topsoil) to use in reclaiming the mined area. In contrast, ores usually constitute a much smaller proportion of the total volume of material removed. Because the excess material expands during processing, there may be more material available to fill the trench than is actually needed. Because of these differences, strip mining of coal and strip mining of ores should probably be regulated differently.

Similarly, the regulatory situation differs for strip mining of iron ore and strip mining of copper ore. The vast majority of iron mined in the United States is consumed in the United States, and nearly all U.S. demand is domestically supplied. Prices for iron are determined primarily by domestic market conditions. In contrast, copper is traded in volatile international markets; domestic producers face stiff competition from foreign sources of supply. A mining regulation that increased costs of both iron and copper mining equally would have very different effects on the two industries. In iron, a few marginal operators might be driven out of the market by the cost increases but the structure of the industry would remain stable. In copper, however, the cost increases caused by the same regulation could place domestic producers at a competitive disadvantage relative to foreign sources, severely reducing the U.S. share of the world market.

There is a general tendency to underestimate the complexity of regulatory problems. Popular discussions and congressional debate often focus on the need for regulation and ignore the difficulties inherent in implementing policy effectively. At the level at which policies are debated, the subtle distinctions among regulated entities that complicate policy design are rarely understood and often overshadowed. To make their case, proponents of new regulation often exaggerate the ease with which government intervention might eliminate a pressing problem. They often stress "need," minimize implementation problems, and concentrate on "horror stories" that are more likely to receive media coverage than detailed discussions. In contrast, opponents of regulatory initiatives take issue predominantly with the declaration of need, because arguing that a proposed regulatory program would be ineffective appears to be a concession on the need issue.[6] Thus, it should not be surprising that the legislative history of the Occupational Safety and Health Act contains numerous references to the thousands of different ways that workers are disabled and little if any consideration of how a single agency with limited resources might formulate policy responses to each of these problems.

It is important to recognize the extent to which regulatory problems consist of many subproblems, because problem diversity affects our choice of regulatory instruments. For example, protecting consumers from product hazards is a fundamentally different regulatory task than protecting people from the radiation hazards of nuclear power. There are only 74 nuclear reactors operating in the United States, and only a handful of substantial technical differences among them. Thus, it is more feasible for the Nuclear Regulatory Commission to prescribe protective measures for each plant than it is for the Consumer Product Safety Commission to identify and eliminate each of the thousands of different hazardous products. Although the NRC may have a more difficult technical task, the CPSC has a considerably more difficult regulatory task. In general, the greater the degree of problem diversity, the less desirable will be centralized regulatory programs that require the government to make complex technical decisions.

Diversity of Regulatees

Regulated institutions differ in many ways. What concerns us about these differences is that they affect the capacity to comply with regulation. Regulators must be sensitive to variation in compliance capabilities for a number of reasons.

First, ignorance of differences in compliance costs may give rise to inefficient allocation of economic resources. For example, if some firms can curb their air pollution more cheaply than others, it is inefficient to mandate uniform reductions in emissions across all firms. By encouraging greater than average reductions in emissions from firms capable of complying cheaply, the same overall reduction in air pollution could be achieved at a lower total cost to society. As a general rule, if the objective of regulation is to influence behavior in the aggregate (such as reducing total particulate emissions, decreasing national energy consumption, or controlling atmospheric release of fluorocarbons), then it will be economically efficient to permit variations in response based on variations in cost of compliance.

Second, severe implementation problems are likely to result if regulators pretend that firms are organized and managed uniformly. For example, a safety regulation that impairs productivity is likely to meet with intense labor opposition in a shop in which workers are compensated on a piece rate, and little or no labor opposition where workers are paid by the hour. Similarly, it is much easier to implement a minority hiring program in a nonunionized firm than in a unionized firm in which job eligibility, promotion, training, and wage scales are governed by seniority provisions.

Third, because regulatees have different compliance capabilities, regulation often affects the competitive structure of an industry in unintended ways. Consider regulation of automobile fuel economy. To encourage energy conservation, Congress has required automobile manufacturers to meet minimum fuel-efficiency standards. To comply, manufacturers have been forced to redesign their cars to save weight. This is an extraordinarily expensive process that taxes the capital reserves of even the largest corporations. Because of its market position and profitability, General Motors has much better access to capital than either Ford or Chrysler. Chrysler has had a particularly difficult time raising the funds necessary to retool, and as a result has been weakened competitively and been placed in a precarious financial position. Furthermore, because it produces fewer cars, Chrysler cannot amortize retooling costs as quickly as its larger competitors. Thus, although apparently neutral with respect to competition, the federal fuel-economy standards have enhanced the market position of the nation's largest automobile manufacturer at the expense of its competitors.

To the extent that firms have different compliance capabilities, regulation will always have competitive consequences (not necessarily undesirable). Uniform regulation will not result in equal treatment. If regulation is to have equal impact along a particular dimension, such as competitive standing, then the design of regulatory policy must take into account differences in compliance capability.

Diversity of Environments

The third way in which diversity complicates the design of regulatory policy is through the setting in which the regulated activity occurs. Variations in environmental conditions can radically alter the nature of the regulated problem. For example, energy conservation is a different problem in buildings located along the coast than in inland buildings, as evidenced recently in Florida. For inland structures, the objective is to maximize the efficiency of air conditioning units by minimizing solar heating of the interior; thus, the recently enacted Florida Energy Code restricts the size of windows and establishes minimum requirements for insulation. But houses along the coast can rely upon ocean breezes for cooling. Consequently, the design objective is exactly the opposite of that for inland homes; they should be open and airy, with large windows and doors. A regulatory strategy for energy conservation in buildings should permit variations in design according to location.

A similar problem arises in regulating water quality. The capacity of a receiving body of water to absorb pollution is a function of tides or currents, the concentration of the pollutant, the location of the discharging source, the temporal distribution of the discharge, the temperatures of both the discharge and the receiving body of water, the cross-sectional area of the receiving body of water, and the natural characteristics of the bottom. Many of these factors change as one travels along the shore of a river or lake. Consequently, the capacity of a receiving body of water to accept a given concentration of pollution will vary depending upon the location of the point source. Thus, even if all pollution sources employ the same technology and have similar abatement cost functions, a uniform discharge regulation may not be desirable, because identical discharges will have different effects on water quality depending on their location.

In general, whenever the nature of a regulated activity is influenced by its location, regulatory policy must take that fact into account.

Conclusions

Regulatory problems are invariably complex and multidimensional. Because of the degree to which circumstances vary among different hazards, regulatees, and environments, a strong case can be made that each regulated institution possesses a slightly different regulatory problem. Thus, there is not just one occupational safety and health problem in the United States; each of the nation's five million workplaces has a different problem. The same observation could just as easily be made about most other regulatory issues.

If job hazards are to be removed from a particular work environment (or if toxic substances are to be controlled, or if energy is to be conserved, or if pollution is to be abated), the regulatory procedures employed must respect unique conditions.

With regard to most regulatory issues—especially job safety and health—the world is too diverse and a regulatory agency's knowledge too limited for it to be able to specify the most effective means for achieving the objective in each regulated institution. Instead, the agency must choose between the promulgation of uniform rules that are likely to work poorly in some situations and the creation of incentives for regulatees to look for ways to abate hazards on their own.

2
"Command and Control" Strategies

Instead of creating incentives for improvements, regulatory agencies in the United States have consistently attempted to mandate solutions to complex problems by what Charles Schultze has termed "command and control" regulation.

OSHA as Example

The Occupational Safety and Health Administration relies upon a system of standards, inspections, and fines to create incentives for making the workplace safe. The Labor Department has primary responsibility for promulgating and enforcing standards.

Most of the OSHA standards currently on the books were adopted during the years immediately following the enactment of the law (1971–1973). During this period Congress permitted the Secretary of Labor to promulgate as OSHA standards "any national consensus standards, and any established federal standards."[1] The intent was to avoid duplicating the efforts of private organizations (such as the American National Standards Institute) that had been in the business of defining safe practices for years. About 4,400 standards were adopted under this procedure. Each standard attempts to address a different workplace hazard. Many of these regulations are very detailed, specifying the exact procedures that must be followed to abate a hazard. For example, ladder safety is covered by over 140 regulations. OSHA has since rescinded many of these consensus standards because they proved to be obsolete.

Scientific input into standard setting is provided by the National Institute for Occupational Safety and Health (NIOSH), an agency of the Department of Health, Education, and Welfare. For each proposed standard NIOSH issues a "criteria document" that summarizes the state of scientific knowledge about the hazard under study. The criteria document also proposes a standard that is based solely on scientific considerations. After receiving the criteria document, OSHA publishes an Advance Notice of Proposed Rulemaking in the Federal Register to solicit public comment. At this point in the process, the OSHA Administrator may appoint a Standards Advisory Committee to make recommendations on the proposed standard. Representatives of labor, management, and academia

13
"Command and Control" Strategies

typically serve on Advisory Committees. The recommendations of the Advisory Committee are considered by OSHA's Office of Standards Development, which is responsible for drafting the final proposed standard. For major standards, a Regulatory Analysis is performed.[2] The proposed standard is published in the Federal Register, comments are solicited, and a public hearing is held. After completion of the hearing-and-comment period, the Agency retires to draft the final standard, which is published in the Federal Register. This process is slow and cumbersome. For example, since its inception OSHA has adopted only 23 new health standards, an average of fewer than 4 per year.

Responsibility for compliance with OSHA standards rests almost exclusively with employers. Although workers have a legal duty to comply with the terms of the act, there are no sanctions for employee noncompliance. Management compliance is monitored through workplace inspections, many of which are targeted at firms in high-risk industries. The remaining inspections are triggered by worker complaints or major accidents. Until 1978, OSHA inspectors performed these inspections by simply showing up at the workplace and demanding entrance, but in that year the Supreme Court ruled that employers could legally refuse to admit OSHA inspectors who failed to produce a search warrant on demand.[3] Because OSHA must demonstrate that probable cause exists in order to obtain a warrant, random searches are no longer permitted.[4]

OSHA's inspection force is small relative to the number of firms subject to regulation: 1,560 federal inspectors are available to inspect the nation's 5 million workplaces. This force is supplemented by an additional 1,500–1,800 state inspectors operating under the aegis of approved state occupational safety and health plans.[5] Because of the size of the inspection force, inspections are infrequent; one study estimates that OSHA is capable of inspecting only 2 percent of the firms it regulates each year.[6]

OSHA inspectors are required by law to issue citations for all standards violations observed. Citations typically result in modest fines. A Senate Oversight Committee reported that from July 1972 through March 1974 98 percent of all OSHA violations cited were minor. The average fine for these violations was $16. Serious violations made up only 1.2 percent of all violations, with fines averaging $648. The remaining violations were classified as "willful," "repeat," or "imminent danger," and fines averaged $1,104.[7]

The regulatory approach mandated by the Occupational Safety and Health Act is striking in light of the number of separate problems that must be addressed as well as the diversity of the regulated institutions. Congress has given OSHA the enormous task of prescribing how every hazard in the nation's workplaces is to be abated. As if this were not a big enough job, OSHA must also write its standards so that they will work well in a wide variety of employment settings. OSHA's job is made even more difficult by the fact that the threats it has at its disposal to encourage compliance—infrequent inspections and modest fines—are unlikely to bring about expensive investments in job safety and health.

In effect, the existing regulatory structure requires OSHA to be omniscient and omnipresent. Congress has enacted a regulatory program that would work well if the occupational safety and health problem consisted of a few major hazards in a handful of homogeneous workplaces. Thus, it should not be surprising that OSHA has been intensely criticized for not promulgating standards quickly enough, for creating standards that do little to improve health and safety conditions and often create more problems than they solve, for imposing excessive costs on employers, and for failing to vigorously protect the health and safety of the American worker. Congress has given OSHA an impossible job, and OSHA has regarded it as possible. The judgment of both institutions is questionable.

Extraoccupational Parallels

The regulatory structure embodied in the Occupational Safety and Health Act is not unique—"command and control" is the dominant regulatory strategy in the United States. This strategy is consistently used in other situations where diversity suggests that centralized regulation is not likely to work well. Rather than create incentives for regulatees to alter their behavior, either by stimulating market mechanisms or by other economic means, we consistently attempt to bureaucratically mandate solutions to complex problems. Not surprisingly, such regulatory programs often are subject to criti-

"Command and Control" Strategies

cisms similar to those leveled at the Occupational Health and Safety Administration's.

The Consumer Product Safety Commission was established in 1973 to protect consumers from hazardous products. This regulatory task is not unlike that of protecting workers from hazards in the workplace. The Consumer Product Safety Act empowers the CPSC to set standards for safe product design and to order the recall or ban the further sale of products that pose "unreasonable risk" to consumers. Standard setting has proved extremely difficult for the CPSC. In 6 years, the agency has adopted only five design standards (governing swimming-pool slides, matchbooks, architectural glass, cellulose insulation, and refuse containers). The CPSC now spends much of its time ordering remedial measures when presented with evidence of unreasonable risks from specific products.

The regulatory structure of the Toxic Substances Control Act of 1975 closely parallels that of the Consumer Product Safety Act. TSCA empowers the Environmental Protection Agency to regulate the manufacture, processing, distribution, use, and disposal of chemical substances and mixtures. If the EPA Administrator concludes that a new substance poses an "unreasonable risk" to health or the environment, he may prohibit the manufacture of the substance or adopt regulations specifying the conditions of its use. Because of the large number of chemicals in use, the difficulty of testing to determine health hazards, and the problems involved in keeping track of substances as they work their way through the industrial process, TSCA imposes immense regulatory burdens on the EPA, which is still struggling to establish priorities for testing and regulating new substances.

Health and safety problems are not the only regulatory issues addressed through "command and control" regulation. Many states are trying to encourage energy conservation by adopting detailed standards for building design. Local building codes regulate the construction process through standards that describe in great detail what materials, processes, and techniques may be used in erecting and modifying buildings. Process standards define what measures must be taken to reclaim strip-mined land. HEW standards regulate the accessibility of public buildings to the handicapped. Congress has even considered (and rejected) a bill that would have required standards to eliminate hazards from summer camps.

What is interesting about all these regulatory programs is that in each case the policy response to a very complex problem was to attempt to mandate a particular solution by promulgating standards that define acceptable conduct. Rarely do we attempt to create incentives for regulatees to alter their behavior on their own. We don't try to abate pollution by taxing it, we don't try to encourage safe product design by improving the incentives of liability law, and we don't try to encourage energy conservation in residential construction by requiring builders to measure and disclose the energy efficiency of the houses they build. When confronted with behavior that we find undesirable, our regulatory response most often consists of bureaucratically defining a very limited range of acceptable behavior and fining regulated institutions that depart from it.

The Political Preference for "Command and Control"

At first glance the choice of standards as the means for regulating hazards appears puzzling. Given the enormous difficulty of setting and enforcing standards for five million different workplaces, one might have expected the authors of the Occupational Safety and Health Act to have looked for other ways to make the workplace safe. Critics of this act have suggested an injury or exposure tax, tort liability, and an enhanced workmen's-compensation system as alternatives to direct regulation. In fact, the political context made the choice of standards almost inevitable.

Like most other legislative draftsmen, the authors of the Occupational Safety and Health Act were not writing on a clean slate. Before the act was adopted 20 states had set up agencies to promote job safety and health. Although these agencies varied in power and effectiveness, they all relied almost exclusively on standards and inspections. The federal government also played a limited role in regulating job safety and health through the Walsh-Healey Public Contracts Act of 1936, which required employers with government contracts in excess of $10,000 to comply with certain health and safety regulations. Walsh-Healey was never very effective, because

17
"Command and Control" Strategies

of a lack of trained inspectors. Many proponents of a comprehensive federal job safety and health bill believed that if they could solve the problems of the state programs (weak mandates and underfunding) and of Walsh-Healey (inadequate inspection), they could solve the job safety and health problem. Thus, they opted for tough standards enforced nationwide by a new agency with a mandate to worry exclusively about health and safety. Alternatives to standards were not even considered. Instead, debate over regulatory structure centered primarily on the issue of whether regulation of job safety and health was better left to the states.

Politicians preferred standards over less direct means of regulation because they gave the appearance of immediate action. While an injury tax might create incentives to remove a particular hazard from the workplace, a standard actually mandated its removal. Moreover, it clearly spelled out the penalty for failure to act: a fine. Standards appealed to politicians because they looked like more tangible evidence of action than a tax scheme or an enhanced workmen's-compensation scheme.

From labor's perspective, standards were preferred because they gave labor an opportunity to participate in what would otherwise be purely management decisions. If labor was to be involved in designing standards, it would have some voice in decisions among alternative production technologies. Similarly, because management's compliance with standards would be mandated by law, labor had the chance to obtain through standards other concessions it could not otherwise obtain at the bargaining table or through less direct methods of regulation (cleaner air, less noise, etc.). Labor also preferred standards to other strategies such as workmen's compensation on the grounds that standards would help the worker before he had an accident. In contrast, labor has argued against accident taxes on the grounds that they would permit an employer to kill or injure workers as long as he was willing to pay the tax.

Finally, as John Mendeloff has observed, business may have been reluctant to suggest an injury tax for fear that "the tax mechanism might metamorphose from a way to prevent injuries into another handy source of public revenue." [8]

While the above analysis may help to understand why the OSH Act relies so heavily on standards, it does not explain why the U.S.

"Command and Control" Strategies

government always seems to use "command and control" strategies in responding to other types of problems. The two most commonly offered explanations are that lawyers, who dominate Congress, are inclined toward regulatory approaches that closely parallel the adjudicatory model, with well-defined rules and sanctions; and that because so few economists serve on congressional staffs economic efficiency is rarely considered in the legislative process. Although there is some truth to both of these theories, they do not explain why direct regulation appeals so strongly to the public as well as to Congress. There appear to be at least four reasons.

First, the political process that culminates in the enactment of a regulatory law often creates a public demand for immediate, visible action. Laws are not passed on impulse. Rather, they are the result of a process of persuasion and bargaining. As James Q. Wilson has noted, during this process proponents of legislation often "[exaggerate] the virtue of those who are to benefit (a defrauded debtor, a sick industry) or the wickedness of those who are to bear the burden (a smog-belching car, a polluting factory, a grasping creditor)."[9] Legislative support is often gained through promises of quick, easy solutions. Thus, standards are adopted because they appear to guarantee through proscription a desired outcome; mandating that particulate emissions from secondary lead smelters shall not exceed a certain level appears to guarantee that they never will. Less direct methods of achieving regulatory goals are overlooked because they permit continuation of the undesirable condition—under an emission tax, the operator of a lead smelter may elect to pay the tax and continue to pollute at pretax levels. It makes little difference to proponents of regulation that standards may be economically inefficient or difficult to enforce. What matters is that standards clearly declare that government will no longer tolerate behavior that the political process defines as undesirable. Standards make a symbolic statement; economic incentives do not.

Second, the concept of a standard has a certain intuitive appeal to it. Even a child understands the logical relationship between a law that proscribes some activity and the activity itself. In contrast, few people who lack formal economic training truly understand how an incentive tax works. Both the logical and the empirical rela-

"Command and Control" Strategies

tionships between the imposition of the tax and the desired outcome seem tenuous.

Third, the public often perceives regulatory problems in a way that leads them to believe that there exists a single "correct" technical solution. Thus, we empower agencies such as the EPA to set standards for drinking water in the belief that men in white coats will be able to define "safe" levels for different pollutants. It matters little that the concept of safety often defies objective definition. What is important is that our notion of a "correct" or "safe" level is inconsistent with a regulatory system based on economic incentives that would allow varying levels of an activity publicly acknowledged to be bad.

Fourth, the public often prefers standards to taxes because taxation schemes are popularly perceived as inflationary and standards are not. In practice, incentive schemes usually impose lower total costs upon the public, although the costs are quite visible. In contrast, a standards system hides the cost of regulation, either by burying it in the price of a product or by converting it into a nonmarket cost. For example, encouraging energy conservation in homes by levying a tax on heating oil imposes visible costs on consumers, while an energy-oriented building code does not.

A possible fifth reason that uniform standards are more popular than economic incentives is that they are perceived as fair. All regulated institutions seem to be treated equally. For example, requiring all electric generating stations to burn low-sulfur coal has a ring of equity. The standard appears to impose equal burdens on all firms. Moreover, it appears to ensure that all people who live near power plants will be exposed to exactly the same health risks. In contrast, if a sulfur-emissions tax is used, some generating stations will find it economical to burn low-sulfur coal exclusively; others will just pay the tax and continue to pollute at pretax levels, and their neighbors may breathe more sulfates. In actuality, popular perceptions about the equity of standards are a bit off the mark. A standard that compels the burning of low-sulfur coal does not impose equal burdens on all firms, because some firms will have readier access to such coal than others. Requiring all plants to burn the same coal does not even standardize the health risks imposed on their neighbors; because of variations in local meteorological conditions and differences in the heights of smokestacks, people downwind of different plants will still be exposed to different risks even if each plant emits exactly the same amount of sulfur. These differences notwithstanding, blunt standards still give the appearance of equity, and the appearance may be more important than equity itself. This preoccupation with appearances leads to what might be termed the paradox of fairness: Any plan to allocate burdens that appears fair must not be fair, because it ignores nonobvious differences among regulatees. For example, odd-even gasoline rationing appears fair until one considers the particular burdens it places on people who are incapable of delaying their purchases of gasoline by one day (such as long-distance travelers, traveling salespeople, and delivery drivers). A corollary to the paradox of fairness is that a truly equitable plan will often be perceived as unfair because the exceptions required to render the plan fair often appear to favor or penalize isolated groups.

Finally, and perhaps most important, the public prefers standards to incentive-based schemes simply because we are used to them. The administrative system is set up to promulgate standards. The legal system is comfortable reviewing standards. Regulatees and interest groups seeking regulation know how to participate in and manipulate a regulatory process that is based on standards. None of these groups has any incentive to change the rules of the regulatory game.

Conclusions

The above analysis does not paint a particularly rosy picture for alternatives to direct regulation. We consistently regulate by government fiat because the political preference for standards is strong. Although growing dissatisfaction with government regulation has produced some movement toward the use of market mechanisms, these legislative efforts have been limited to economic regulation (specifically, regulation of the airline and trucking industries).

The public seems slightly more comfortable relying on indirect incentives when the object is to hold down prices rather than to abate pollution or eliminate job hazards, for two reasons. People have a much better understanding of the relationship between

"Command and Control" Strategies

competition and prices, which they can observe daily in the marketplace, than of the relationship between a sulfur-emissions tax and smog. Also, while people may believe that there exists a safe level for exposure to food additives, they do not generally harbor a similar belief in an objectively correct price—the public's faith in scientists is much greater than its faith in economists.

Critics of regulatory policy need to recognize and understand political forces that favor standards, because the design of regulatory policy is constrained by political realities. If regulatory reform is to be more than the rallying cry of academic economists, we must be prepared to look for ways to design policies that not only are efficient, but also respond to political demands.

II
EVALUATING POLICY

3
The Impact of OSHA

Evaluating the impact of the Occupational Safety and Health Act is difficult for a number of reasons. First, as with many evaluation efforts, the data are poor. The fact that section 8(c)3 of the OSH Act changed the manner of reporting accident and illness data frustrates pre- and post-OSHA comparisons. Second, the numbers are small. This makes it difficult to ascertain small changes in accident rates with confidence. Third, factors other than OSHA influence accident and illness rates (for example, the rate of hiring, the amount of overtime worked, and the level of capacity utilization). And fourth, OSHA just has not been around long enough to have had clearly measurable effects. Most occupational diseases result from cumulative lifetime exposure to hazards such as cotton or coal dust. Thus, a reduction in exposure due to a program like OSHA is unlikely to have a measurable impact on disability from occupational disease for years to come.

The above considerations suggest that we are more likely to find an impact on the accident rate than on the illness rate. This conclusion is buttressed by OSHA's intense concentration of its enforcement resources on the abatement of safety hazards during the first 5 years of its existence. A number of analysts have evaluated OSHA safety programs using multiple regression techniques to control for external factors. While their work is constrained by the data, it nonetheless illustrates the likely magnitude of OSHA's impact on deaths and injuries from job-related accidents.

Injuries to Employees

John Mendeloff has estimated the rate of change in lost-time injury rates in manufacturing in California, using data compiled under the old reporting system from 1948 to 1970 and post-OSHA data through 1974.[1] Mendeloff's model makes the reasonable assumption that the annual rate of change in lost-time injury rates is structurally similar under the new reporting system to what it was under the old. Thus, he is able to estimate what the rate of change in manufacturing lost-time injury rates would probably be in the absence of OSHA. Mendeloff estimates this dependent variable as a function of the hiring rate, the hiring rate lagged 1 year, the percentage of workers aged 18–24, and the average hourly wage of production workers. The estimated coefficients are then used to predict changes in the injury rates for post-OSHA years 1972–1973 and 1973–1974 and to attribute to the effects of OSHA the difference between predicted and actual changes, as shown in table 3.1. Mendeloff's model predicts that the injury rate rose 0.8% less in 1972–1973 and 2.2% less in 1973–1974 than it would have in the absence of OSHA. These results, however, are not statistically significant. The changes may reflect random variations in the data. The same type of estimation using California workmen's compensation data yields similarly inconclusive results.

Mendeloff also applied his model to two specific types of accidents that a panel of safety engineers felt most likely to be prevented by a system of standards and inspection.[2] These accidents are the "caught in and between" type, in which a worker is injured through contact with moving parts of machinery, and the "falls and slips" type. Mendeloff concludes that the incidence of these accidents under OSHA is significantly less than predicted in 1974. From these results he estimates that OSHA may have reduced California's manufacturing injury rate by 2–3 percent and its work-related fatalities by as much as 5 percent. It is not clear to what extent these conclusions can be generalized. California had a relatively strong state occupational safety and health program prior to the implementation of OSHA in 1971. OSHA preempted this program until 1974 when California began conducting inspections under a federally approved state plan, greatly increasing both the probability of inspection and citation over the national average. Presumably, the strong pre-OSHA state program biases the results downward relative to OSHA's impact on a nationwide scale. The strong post-OSHA enforcement should have the opposite effect. Unfortunately, it is not possible to determine which effect is larger.

Table 3.1
Change in Accident Rates Attributable to OSHA

Year	Predicted Change in Injury Rate Without OSHA	Actual Change in Injury Rate	Change Attributable to OSHA
1972–1973	+ 7.9%	+ 7.1%	− 0.8%
1973–1974	+ 6.6%	+ 4.4%	− 2.2%

Robert Smith of Cornell estimated a cross-sectional log-linear model of 109 three-digit Standard Industrial Classification manufacturing industries, using both 1972 and 1973 national injury rates as dependent variables.[3] The purpose of Smith's analysis was to test whether industries that have been singled out by OSHA for more frequent inspections (under the Target Industry Program) have exhibited lower injury rates. Smith's independent variables include the 1968, 1969, and 1970 injury rates, the proportional change in industry employment, and a dummy variable indicating whether the industry is a member of the Target Industry Program. Using several variations of the above model, Smith found that the coefficient of the dummy variable is positive, which suggests that target industries have higher injury rates than expected (although the coefficient of the dummy is not significant). One possible explanation for this is that the Target Industry Program has improved accident-reporting procedures in affected industries. More recent analysis indicates that the Target Industry Program does have the expected downward influence on injury rates, although the results are still not statistically significant.

Aldonna Di Pietro analyzed the impact of OSHA inspections on the injury rates of individual firms to determine whether inspected firms exhibit significantly lower injury rates in the year following inspection.[4] She formulated a model that explains the 1973 firm injury rate as a function of the 1972 injury rate, the proportional change in its level of employment, and whether the firm was inspected in 1973. Separate estimates were made for small, medium, and large groups of firms within each of 18 different two-digit SIC classifications. Di Pietro concluded that in most industry groups the data do not show any statistically significant reductions in injury rates of inspected firms. Moreover, where the inspection dummy is significant, it is far more likely to be positive than negative. Di Pietro explained this apparent contradiction by observing that inspectors are likely to have concentrated their efforts on accident-prone firms. This would account for the positive coefficient for the inspection dummy.

Kip Viscusi uses pooled time series and cross-sectional data from 61 two-digit industries for the years 1972–1975 to estimate the impact of OSHA on industry injury rates and industry capital investments in health and safety.[5] Industry inspection rates as well as the proposed OSHA penalty rates enter into this model as independent policy variables. Viscusi used the industry percentage of female workers and production workers as proxies for low- and high-risk jobs among industries. Age mixture and racial mixture of workers were also controlled. Cyclical variations were accounted for by incorporating the percentage change in employment, hours worked per week, and weekly overtime into the model. Viscusi concluded that OSHA has had no demonstrable effect on industry health and safety investments or industry injury rates. The investment conclusion is interesting because it appears to rebut the claims of business that OSHA has mandated large additional capital expenditures. Viscusi notes that the coefficients of the annual dummy variables for the investment model also are not significant, which suggests that OSHA has not even had an indirect effect on investment patterns. Health and safety investments appear to be motivated by the nature of an industry's technology and the makeup of its industry. The annual dummy variables *are* significant in the injury-rate model, a fact that Viscusi found difficult to explain. He attributed these time effects to a statistical oddity in the sample, to growing laxity in injury reporting over time, to continuation of a long-run trend, or to a possible (and erroneous) public perception that OSHA is vigorously enforcing standards. A more charitable explanation might be that OSHA has stimulated parallel and effective health and safety activity among workers, and that these have reduced the annual injury rate.

A rough comparison of 1974 and 1975 injury statistics suggests that OSHA's impact may be confined to less severe accidents. The Bureau of Labor Statistics reports that the evidence of lost-workday *cases* (that is, injuries and illnesses that resulted in lost workdays) per 100 full-time workers declined from 3.5 in 1974 to 3.3 in 1975.[6] However, the number of lost *workdays* per 100 workers increased from 54.6 to 56.1 during the same period (see table 3.2). In other words, there were fewer cases in 1975 than in 1974, but the cases were more severe. There are a number of possible reasons for this phenomenon. First, as noted above, federal regulatory efforts may be more effective in reducing accidents with small consequences than in reducing accidents that lead to loss of workdays. Second,

Table 3.2
Loss of Workdays After Nonfatal Job-Related Injuries and Illnesses per 100 Full-Time Workers

Year	Total Cases	Lost-Workday Cases	Cases Without Lost Workdays	Lost Workdays
1974	10.4	3.5	6.9	54.6
1975	9.1	3.3	5.8	56.1

Source: Bureau of Labor Statistics, U.S. Department of Labor.

some accidents of a type categorized as not leading to loss of workdays in 1974 may have been classified as lost-time accidents in 1975. A shift of a few accidents between categories would account for the apparent increase in lost workdays. Third, factors such as capacity utilization, hiring rate, and amount of overtime worked may not affect severe accidents to the same degree as they affect less severe accidents. If this assumption is true, the data might be explained by differential injury-rate responses to changes in causative factors.

In many respects, the above analyses shed more light on the difficulties of evaluating OSHA than on the actual performance of the agency. A sympathetic reading would suggest that the agency is too young and the data too "soft" to allow judgment of the actual impact of OSHA on job-related accidents. On the other hand, it is also possible to reach the opposite conclusion: Because the OSH Act focuses on permanently alterable characteristics of the workplace (machine guarding, shoring of trenches, grounding of electrical circuits, etc.), it should have had its greatest impact in its first years, as permanent hazards were abated. According to this theory there would have been a one-time drop in the accident rate as firms came into compliance with the act. The analyses described above do not conclusively establish that this one-time drop has not occurred or will not occur, but by focusing on those areas where OSHA regulation could be expected to have its greatest effect (the Target Industry Program and the accident rate of inspected firms) they offer reasonably persuasive evidence that OSHA has had no large measurable impact on the rate of accidents in the workplace. Moreover, it is possible that whatever impact OSHA has had on accidents is confined to less serious accidents that involve little loss of work time.

Costs to Employers

OSHA imposes costs on regulatees in three ways. First, businesses are fined for noncompliance with OSHA regulations. As noted earlier, the fines are relatively modest; the total amount collected in 1975 was $9.5 million.[7] In an economic sense, fines alone impose no cost on society, since the funds are only transferred from the business sector to the general treasury. Second, regulatees also bear the real costs of purchasing and maintaining health and safety technology. These capital investments, which are usually required to bring firms into compliance with OSHA standards, represent real costs to society because they represent consumption of scarce resources that would otherwise be used in the production of other goods and services. Third, health and safety regulations often impair worker productivity. For example, the Department of Labor's Inflationary Impact Statement on the proposed coke-oven emission standards estimates that coke production per worker may decline by as much as 29 percent as a result of changes in work practices mandated by the standard.[8] Labor productivity declines also impose real costs on society.

Three sources of data exist on the cost of compliance with OSHA regulations. Since 1973 McGraw-Hill has conducted a survey of the cost of capital investments in health and safety by industry.[9] Shortly after the passage of the OSH Act, the National Association of Manufacturers also surveyed its members to determine the likely cost of compliance in relation to firm size.[10] Also, in response to a presidential order OSHA has commissioned a number of Inflationary Impact Statements for proposed standards.[11] The McGraw-Hill and NAM survey results are summarized in table 3.3.

According to the McGraw-Hill data, capital expenditures for occupational safety and health have averaged slightly over $3 billion per year since 1972, divided roughly equally between manufacturing and nonmanufacturing. But, as noted earlier by Viscusi, it is not clear what proportion of these expenditures actually is attributable to OSHA. The NAM estimates that firms with fewer than 100 employees will spend on the average $35,000 to comply with OSHA and that those with more than 5,000 employees will spend close to $4.7 million each. The total amount predicted to be spent for compliance is approximately 3 percent of all capital expenditures by firms. It must be stressed, however, that these are predicted expenditures, not actual expenditures.

Table 3.4 summarizes the estimated industry cost of compliance for four proposed OSHA standards. These estimates illustrate the likely *additional* costs of compliance with new regulations. The noise standard is likely to have the most far-reaching impact of any standard to date, because it applies to all sectors of the economy. The coke-oven standard is significant because it mandates major retooling and redesign of existing coke-oven technology. A closer look at the cost of the proposed coke-oven standard yields some interesting insights into the types of tradeoffs that are foreclosed by OSHA regulation. The low estimate for the annual cost of compliance is equivalent to an annual expenditure of $8,130 per exposed coke-oven worker; the high estimate works out to an expenditure of $43,000 per exposed worker per year.[12] (Since the employer's duty to comply is owed to the government, OSHA regulation does not permit workers to waive their rights under the OSH Act in return for wage premiums; this precludes some likely Pareto-optimal exchanges.)

While the above estimates give some feel for the magnitude of the cost of OSHA regulation, they are far from conclusive. Measurement of the costs of both capital investments and productivity impairment attributable to OSHA is imprecise at best. Estimation of these costs by survey suffers from the natural inclination of businessmen to exaggerate the cost of compliance with government regulations. Nonetheless, there does appear to be some evidence to support the claims of critics that compliance with OSHA is very costly. Moreover, given OSHA's probable future emphasis on health, compliance costs may increase substantially.

So far, this chapter has presented reasonably strong evidence that OSHA has yet to demonstrate significant impact on the rate of occupational accidents. This conclusion is disturbing in light of the high costs OSHA imposes and in light of the agency's intense concentration on the abatement of safety hazards. In the area where OSHA's impact should be the strongest, we are unable to detect any significant changes in outcomes. In the remainder of this chapter we will attempt to explain why OSHA has not been more suc-

The Impact of OSHA

Table 3.3
Cost of Compliance with OSHA Regulations

McGraw-Hill Survey of Capital Costs of
Occupational Safety and Health
(in Millions of Dollars)

Industry	1972	1973	1974	1975	1978
Manufacturing	938	1,207	1,577	1,638	1,847
Nonmanufacturing	1,157	1,362	1,497	1,479	1,863
Total	3,509	2,569	3,074	3,117	3,710

National Association of Manufacturers
Estimated Cost of Compliance with OSHA by Firm Size
(in Dollars)

No. of Employees	Safety	Health	Total
1–100	24,000	11,000	35,000
101–500	50,500	23,000	73,500
501–1000	141,140	209,627	350,767
1001–2000	272,000	58,630	330,630
2001–5000	552,000	278,000	830,000
5000–	2,226,000	2,455,000	4,681,000

Table 3.4
Costs of Compliance with Proposed OSHA Regulations

Proposed Standard	Total Expected Capital Cost	Annualized Cost of Compliance
Noise		
90 dBA	$ 10.5 billion	$ 1.58 billion
85 dBA	$ 18.5 billion	$ 2.78 billion
Inorganic arsenic		
0.1 mg/m^3	$ 18.4 million	
0.05 mg/m^3	$ 55.4 million	
0.004 mg/m^3	$110.8 million	
Coke-oven emissions		$240.6–280 million
Commercial diving		$ 22 million

Source: OSHA Inflationary Impact Statement

cessful in controlling the rate of occupational accidents. This chapter will also demonstrate that even with good management and perfect compliance OSHA's ability to curb job-related accidents would be very limited.

A Model of Hazard Abatement

Three separate events or processes must occur if a hazard is to be removed from the workplace: The hazard must be identified as a hazard, an effective strategy for its abatement must be planned, and the plan must be implemented. Failure of any of these events results in failure to abate the hazard. Figure 3.1 illustrates this model.

Figure 3.1
A Model of Hazard Abatement

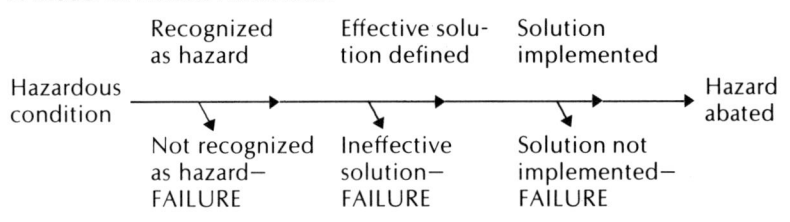

```
                    Recognized         Effective solu-    Solution
                    as hazard          tion defined       implemented
Hazardous    ─────────────────▶───────────────────▶──────────────────▶  Hazard
condition           ↘                  ↘                   ↘            abated
                    Not recognized     Ineffective         Solution not
                    as hazard—         solution—           implemented—
                    FAILURE            FAILURE             FAILURE
```

Identification of Hazards

If any hazard is to be removed from the workplace, it must first be identified as a hazard. This is a nontrivial task, especially in the case of health hazards and low-probability safety hazards. How well an abatement process correctly recognizes hazards is clearly a factor in its effectiveness. Hazards may be identified in any of the following ways:

observation of actual accidents, near misses, and illnesses,

workplace-specific review of conditions by hazard-conscious people, such as safety officers, industrial hygienists, and safety committees,

attention to demands of workers for risk premiums for performing certain operations, or

reference to journals, trade-association publications, regulations, and other sources that report formal academic research and information gained from accident experience at other workplaces.

None of these identification procedures will reveal every hazard. For example, observation will be relatively inefficient in identifying carcinogenic chemicals with long latency periods, and published information is unlikely to contribute to identification of workplace-specific hazards.

Definition of Solutions
After a hazard is identified, the level of risk it poses to workers must be determined. Three factors must be considered: the probability of an undesirable outcome if the hazard is left unabated, the number of workers exposed to the risk, and the extent of the harm in the worst case. These calculations are necessary to reach a decision on the resources, if any, to be committed to abatement.[13] In defining a solution, alternative abatement strategies must be reviewed and evaluated for cost and effectiveness; then one must be selected for implementation. Abatement strategies include the following:

engineering solutions, that is, redesign or modification of equipment or the workplace to eliminate the hazard (examples are installation of machine guarding, elimination of toxic chemicals at the source, and grounding of electrical circuits),

use of protective devices, such as earplugs, safety glasses, respirators, and wire-mesh gloves,

worker-oriented solutions, such as programs to train workers in safety practices and to teach them to recognize hazards, and

administrative solutions, that is, restructuring of work practices and routines to abate or limit exposure to hazards (examples are job rotation, changes in speed of production processes, contracting out of particularly hazardous operations to businesses specifically trained and equipped to perform them safely, and use of backup personnel to assist workers performing hazardous operations).

There are also strategies for containing the impact of an occupational accident or illness once it occurs. For example, some workplaces have in-house infirmaries, plant doctors, plant ambulances, and first-aid equipment.

Implementation of Solutions
Hazard-abatement equipment must be purchased and installed, and workers must be trained to use it. If necessary, work practices and assignments have to be adjusted. A monitoring system should

be established to ensure that the hazard-limiting solution is being implemented as specified; ideally, this monitoring system should also gather information to permit continual evaluation of the effectiveness of the abatement procedures.

The problems inherent in implementing a plantwide occupational safety and health program are formidable. In the short run, most hazard-abatement strategies impair productivity. As a result, a firm's profit and production goals will often be at odds with its stated health and safety objectives. Although it is possible for top management to resolve this conflict, effective implementation of a hazard-abatement program requires that the means of resolution be communicated to the middle- and lower-level managers who will oversee the implementation. When production requirements are not revised in consideration of expected productivity losses, supervisors are likely to slight implementation of health and safety programs. Given a choice between pushing health and safety (traditionally a low-priority item for both labor and management) and pushing production, supervisory personnel nearly always choose production unless it is made clear to them that they should do otherwise. Getting middle- and lower-level supervisors to take occupational health and safety very seriously—something most have never done—is essential to effective implementation.

Hazard-abatement strategies also interact with other work rules and conditions of employment. For example, protective gloves may slow down assembly workers, thus necessitating renegotiation of the piece rate. Elimination of hazards may upset salary structures if some workers had been receiving wage premiums for performing hazardous work. Reordering or redesign of production processes often entails redefinition of job assignments. Some hazard-abatement approaches, such as automation or contracting out of hazardous work, threaten job security. These secondary impacts create significant problems for implementation. Health and safety programs can upset work rules governing terms and conditions of employment that are the fruits of a lengthy, subtle, and complex process of negotiation and compromise between labor and management. The resulting jurisdictional and wage disputes are likely to be costly to resolve. Consequently, the second step in hazard abatement—definition of solution—must take into account the likely costs of implementation. When the definition of the solution cannot be made to preclude these secondary impacts, the implementation process must include means of ameliorating them.

OSHA Regulation Evaluated in Terms of the Hazard-Abatement Model

Identification of Hazards

The present regulatory structure relies solely upon information transfer to facilitate hazard identification. OSHA collects and evaluates information about unsafe working conditions and disseminates it by promulgating regulations that signal the existence of hazards to employers. This is the most centralized method of hazard identification possible. The present system creates few if any incentives for employers to identify and abate hazards not covered by OSHA regulations.[14] Consequently, to the extent that hazards are not identified by OSHA (or by market forces) they will not be identified or abated at all. There are strong theoretical reasons for relying upon centralized procedures for identifying health hazards. However, the case for identifying safety hazards by the same process is far less compelling.

Health-hazard research is costly. It requires a large data base, technical sophistication, and sufficient resources to sustain long-term studies. Because the information produced is a public good, it is unlikely that enough of this research would occur without government intervention. Individual employers have little incentive to incur the large costs necessary to produce useful information because the potential financial return to any one employer—a diminished wage bill and reduced insurance premiums—is small. Because firms cannot easily and cheaply pool their research resources, they can support little basic research. Thus, there is a strong case for having the government do the basic research necessary for identifying occupational health hazards.

It is far more difficult to justify centralized recognition and identification of safety hazards on either theoretical or practical grounds. In theory, safety hazards have a greater tendency than health hazards to be unique to individual workplaces. For example, the major health hazard in the asbestos industry is asbestos dust, which is thought to cause lung cancer, pleural cancer, and asbestosis (scarring of lung tissue). It is a problem in virtually all phases of asbestos production, fabrication, and installation. Moreover, the nature of the problem is similar in all plants: When handled for any purpose, asbestos produces a fine dust that can be inhaled by workers. Because of the universal nature of the asbestos problem, the information produced by asbestos research is a public good. In contrast, safety problems vary both in kind and in degree from place to place. For example, inadequate shoring and trenching may be a major problem in one mine; incorrectly maintained jackhammers may be a problem in another. If hazards are workplace-specific rather than industrywide, private markets are more likely than centralized regulation to encourage efficient investment in hazard identification (if other incentives for hazard abatement are present). Like the hazards, the information produced from investment in safety-hazard identification is workplace-specific. If there is no demand for this information outside the firm producing it, then clearly the information is not a public good. There still may be good reasons, however, for government to encourage employers to take additional steps to identify safety hazards. Inadequate risk premiums, wage problems, bargaining imbalances, and external factors may all dull the incentives for employers to invest in job safety, and may justify some forms of intervention. Employers may not be doing enough to identify and abate safety hazards, but not because safety information is an undersupplied public good. More safety-hazard identification may be needed, but there is little theoretical rationale for supplying it collectively.

There are also strong practical reasons for not centralizing responsibility for safety-hazard identification as the current regulatory structure does. Hazards differ across many dimensions, and a government agency like OSHA or NIOSH is only capable of recognizing a small proportion of them. For example, as noted above, many hazards are unique to individual work environments. Without vast investigative resources, a centralized regulatory authority will concentrate on identifying the unsafe conditions common to all workplaces. Specific hazards will go undetected and unabated. Similarly, new hazards created by new technology will not be recognized quickly by a regulatory structure that takes years to promulgate a standard. OSHA's preference for abating hazards through engineering controls (the easiest strategy to enforce through inspection) also biases its standard setting and hazard investigation in favor of environmental hazards. Hazards created by careless, inattentive, or inadequately trained workers are given little attention because these hazards are difficult to abate by standard setting and inspection. Hazards created by momentary unsafe conditions are also difficult for a centralized regulatory authority to identify. In sum, the current regulatory structure concentrates on the identification of well-known hazards, common to many workplaces, that are permanent, alterable characteristics of the environment. New, temporary, workplace-specific, and work-related hazards are unlikely to be identified by OSHA. As a result, occupational accidents caused by these conditions are unlikely to be prevented by the current regulatory structure, even with perfect enforcement and compliance.

We can assess the potential impact of OSHA on the total rate of job-related accidents by estimating the relative numbers of worker-related and workplace-related hazards. A number of analysts have attempted to measure the proportion of accidents attributable to such factors as carelessness, inattentiveness, and lack of experience—factors that are particularly difficult to control with physical standards. The results of six such studies are summarized in table 3.5.

The large variation in estimates of the proportion of accidents attributable to unsafe acts is probably due to methodological problems in categorizing the causes of accidents. The unsafe-acts figure is very sensitive to the algorithm used to distinguish accidents caused exclusively by unsafe acts and those caused by the interaction of acts and conditions. This type of analysis also must contend with biases in reporting. The injury reports typically used to evaluate the causes of accidents are usually filed by employers, who might be expected to underreport unsafe working conditions. Sample-selection bias may also affect the distribution of causes. For example, a sample drawn from a labor-intensive industry might be expected to exhibit a higher proportion of worker-related accidents than a similar sample drawn from a capital-intensive industry.

Notwithstanding these problems, the information provided in table 3.5 is of some value. There is consensus that worker-related

Table 3.5
Estimated Percentages of Accidents Due to Unsafe Acts versus Unsafe Conditions

Study	% Due to Unsafe Acts	% Due to Unsafe Conditions	% Due to Combination	% Unknown
Pennsylvania Dept. of Labor*	2%	3%	95%	
National Safety Council*	19	18	63	
Mintz and Blum†	21		79	
Hagglund†	35	54	4	7%
Heinrich*	88	12		
Furniss*		16	84	

* Cited in National Commission on State Workman's Compensation Laws, *Compendium on Workman's Compensation* (Washington, D.C., 1972), pp. 287–288.

† Cited in N. Ashford, *Crisis in the Workplace* (Cambridge, Mass.: MIT Press, 1976), pp. 109–114.

factors are involved, at least in part, in a large proportion of accidents. Estimates of this figure (the sum of the "acts" and "combination" columns in table 3.5) range from 39 percent to 97 percent. The Hagglund study casts the present regulatory structure in the best light. However, it estimates that only 54 percent of all accidents are caused exclusively by the hazards most readily identified and abated under OSHA: environmental conditions. The other estimates cluster just under 20 percent. Although these numbers are far from certain, they suggest that it is at least likely that the present regulatory approach has a very limited potential for affecting the rate of job-related accidents, because many accidents are not caused by environmental factors.

Two studies in which accident reports were reviewed to determine whether injuries could have been prevented by perfect compliance with existing safety regulations offer a slightly better indication of the potential contribution of OSHA to hazard identification and abatement than those cited in table 3.5. By focusing entirely on whether compliance with existing safety regulations would have prevented injury, they avoid the difficult methodological problem

The Impact of OSHA

of distinguishing accidents according to cause. The results of these two studies are summarized in table 3.6. (It should be noted that the safety codes in effect in Wisconsin and New York at the time of these studies were not identical to current OSHA regulations. However, as OSHA simply adopted preexisting national-consensus standards, data gathered under the old Wisconsin and New York codes should be relevant to the current OSHA safety standards.)

The higher incidence of code violations in serious accidents appears inconsistent with the evidence presented earlier in this chapter that OSHA's impact may be confined to less serious injuries. It is also possible that the researchers looked harder at serious accidents to discover code violations. Nonuniform enforcement of OSHA regulations may also explain this discrepancy. There is evidence that during the first 6 years of OSHA the agency's inspectors concentrated on citing violations of minor regulations.[15] Thus, OSHA may have had more impact on minor accidents, although its potential comparative advantage may be in the identification and abatement of more serious hazards. In any event, the analyses cited in table 3.6 suggest that a centralized standards-oriented regulatory approach is at best capable of affecting only 30 percent of nonserious accidents and 57 percent of serious accidents. These results are consistent with the data presented in table 3.5.

The ability of inspectors to observe violations of safety regulations also constrains the potential contribution of OSHA to hazard identification and abatement. Under the present regulatory structure, the primary incentive for compliance with regulations is the

Table 3.6
Proportion of Accidents Involving Code Violations

Study	Number of Cases	% with Violations
Wisconsin*	257 (90 fatal)	30 (39)
New York†	3,216 (206 serious)	22.4 (57)

* Cited in Ashford, p. 114.

† Cited in W. Oi, "On Evaluating the Effectiveness of the OSHA Inspection Program," U.S. Department of Labor contract L-72-86, May 1975, p. 42.

threat of inspection and fines. This threat is least effective where hazards and violations are not readily observable by infrequent outside inspection. Momentary dangerous conditions, for example, are not observable unless present at the time of inspection. Similarly, hazards created by inadequate or faulty maintenance typically are recognizable only by people familiar with the technology and work routines of the particular workplace. Mendeloff reports that a panel of California Division of Industrial Safety engineers concluded that only 18.4 percent of 920 injuries studied might have been prevented if an inspector had visited the plant shortly before the day of the injury, and that an additional 8 percent might have been prevented by in-house inspection.[16] A Wisconsin Department of Labor report estimated that 25 percent of accidents reviewed could have been prevented by the abatement of physical hazards subject to inspection, and attributed 45 percent of the accidents to acts by workers and 30 percent to transitory hazards.[17]

In summary: OSHA has yet to affect the rate of job-related accidents, because many of these accidents are structurally immune to OSHA-type regulation; OSHA is incapable of defining regulations that identify workplace-specific hazards, hazards created by workers themselves, and momentary unsafe conditions; and some hazards cannot be detected by infrequent outside inspection. Even with perfect compliance, OSHA may be capable of preventing no more than a quarter of all occupational accidents.

Definition of Solutions
OSHA regulations do more than signal the existence of hazards to employers and employees. They also specify the maximum level of risk to which workers may legally be exposed. Some standards also go one step further and mandate specific methods of hazard abatement. The process of solution definition that has evolved under OSHA is complex and not fully understood. Decisions and judgments about risks and methods of compliance are affected by scientific, technical, political, bureaucratic, legal, and economic considerations.

Standard setting begins with an assessment by NIOSH of the scientific knowledge about a particular job hazard. NIOSH's research culminates in a "criteria document," which contains a proposed standard and documents the biological effects of exposure to the hazard. The criteria document is based primarily on scientific and technical considerations, with minimal input by either labor or management. NIOSH has no economic-analysis capability of its own.

Upon receipt of the criteria document, OSHA publishes an Advance Notice of Proposed Rulemaking in the Federal Register, soliciting public comment. The OSHA Administrator may appoint a Standards Advisory Committee to make recommendations on the proposed standard. The typical committee consists of fifteen people from labor, management, government, and the public, selected for their interest and expertise in the matter under consideration. The committee is constrained by its lack of a full-time staff and by a statutory requirement that it produce a recommendation within 270 days.

The Standards Advisory Committee represents the first real opportunity for the presentation of parochial perspectives in the standard-setting process. The committee may receive testimony in which labor and management representatives present alternative proposals for standards. The NIOSH criteria document is also considered in this process. Often these proposals are a world apart, reflecting fundamental philosophical differences between labor and management. Labor usually takes the position that costs are not a relevant concern where the lives and well-being of workers are at stake, and all known hazards should be eliminated from the workplace. Management is typically reluctant to spend large sums to comply with regulations of dubious efficacy, and often argues that some risks are so small that expensive abatement efforts simply are not justified.

Although the Standards Advisory Committee provides a convenient forum for alternative viewpoints, it is not well structured to resolve these differences. The lack of investigatory resources prevents serious objective analysis of either the extent of the risks or the magnitude of the costs, benefits, and implementation problems associated with different proposed standards. Labor cites one set of numbers and management another; few minds are changed. Consensus is also inhibited by the absence of any pressure to reach agreement. In contrast with contract-bargaining negotiations, committee meetings attach no cost to nonagreement. Even if the committee reaches agreement, the decision is not binding on anyone, since committee members usually lack the power to bind their principal organizations. Consequently, repudiation of a Standards Advisory Committee's recommendations by the represented organizations is not uncommon. Disputes over proposed standards are resolved by a simple majority vote that often reflects the *a priori* positions of the parties. Rarely, if ever, is lasting consensus achieved.

The recommendations of the Standards Advisory Committee are only advisory and may be rejected by OSHA's Office of Standards Development, which is responsible for drafting the final proposed standard. Scientific and technical considerations dominate OSHA's review of the Standards Advisory Committee's recommendations. Labor and management have no input in this process. The draft of the proposed standard is reviewed by OSHA's Solicitor Office, Office of Information, Office of Regional Programs, and Assistant Secretary before publication of the proposed standard in the Federal Register. For regulations that may have major economic consequences, Executive Order 12044 requires the preparation of a "Regulatory Analysis" that describes the alternatives considered by the agency, the economic consequences of each alternative, and a detailed explanation of the reasons for choosing the preferred alternative. The draft Regulatory Analysis is published at the same time as the proposed standard. After publication, comments are solicited from interested parties and other government agencies. The Regulatory Analysis represents the first serious attempt in the standard-setting process to assess the likely costs of the proposed regulations. Both the merits of the proposed standard and the cost estimates contained in the Regulatory Analysis are debated at a public hearing held shortly after publication. The positions taken by people participating at the hearing are predictable. Labor typically offers evidence supporting the most stringent standard that is scientifically defensible. Management representatives counter with their own experts, who invariably minimize the risks and emphasize the costs and difficulties associated with compliance. In theory, the purpose of the hearing is to gather additional information and provide a forum for public participation in the standard-setting process. In practice, much of the testimony is offered only to build a record for future legal appeal of an adverse OSHA ruling.

After the hearing, OSHA goes to work drafting the final regulations. This is supposed to be the job of the project director in the Office of Standards Development, but in fact the actual decision concerning both the degree of protection to be required by the standard and the permissible methods of hazard abatement may be made by anyone in the OSHA hierarchy from the project director up to and including the Assistant Secretary. Whatever decision is made represents OSHA's best effort to resolve the differences of labor and management expressed in the Standards Advisory Committee and the public hearings. The criteria used by OSHA to determine the degree of protection to be required by a standard are essentially scientific. OSHA has adopted the position that where no absolutely safe exposure level or safe practice exists—as is the case for nearly all health hazards and many safety hazards—the standard should be as protective of workers as technology and the financial condition of the industry will permit. Costs are considered relevant only in determining how much health and safety an industry can afford without driving a significant number of firms out of business,[18] and sometimes as a factor in how compliance is to be phased. This is an implicit rejection of the management position that some reductions in risk are not justified in light of costs. (Management is currently contesting OSHA's position on costs in the courts.)

There are a number of criteria that could be used to evaluate the process just described. An economist would focus on whether the process produces outcomes that are globally efficient, whether the standards encourage proper allocation of society's resources to the production of occupational safety and health. In an economic sense, resources are misallocated if the benefits gained from inframarginal investments in hazard abatement are exceeded by the costs. Since OSHA makes no attempt to assess the likely benefits from standards compliance, let alone assign values to them,[19] it is very unlikely that the present process produces globally efficient outcomes.

It can be argued that global efficiency is not a relevant criterion for assessing job safety and health policies. Many people find it morally offensive to determine the allocation of resources to hazard abatement on the basis of an explicit monetary valuation of human life. Furthermore, explicit valuation of life may be politically unacceptable. However, even if we decline to assign a value to human life and instead determine the resources to be used to make the workplace safe by some other process (such as a political or a moral process, or one based on equity), we should still attempt to maximize the return on our investment in job safety and health. In other words, our process should save as many lives, avert as many accidents, and avoid as many cases of illness as is possible for a given commitment of resources to hazard abatement. The OSHA standard-setting process does not satisfy this cost-effectiveness criterion for three reasons.

First, as noted above, OSHA makes no attempt to determine how many deaths or injuries might be prevented by different standards, thus precluding comparison of alternatives on the basis of avoided morbidity or mortality.

Second, OSHA standards mandate uniform performance for each firm affected by a standard. All firms must comply with all relevant standards equally. Although this approach appears equitable and may be politically compelling, it ignores the diversities described in chapter 1. If the cost of compliance differs from firm to firm, then it may be possible to save more lives for a given compliance expenditure by tightening standards for firms that can save lives cheaply and loosening them for firms that can only eliminate hazards at great expense.[20]

Third, many OSHA standards mandate specific methods of hazard abatement. For example, the original punch-press standard (since modified) required an automatic feed of the metal stock into the press to eliminate the possibility that a worker might activate the press while his hands were positioning the material in the stamping area. Specification standards prevent firms from using—and thus discourage them from developing—least-cost methods of abating hazards.[21]

In addition to global efficiency and cost-effectiveness, an economist might also ask whether the current standard-setting process produces Pareto-optimal outcomes from the perspective of any individual firm. Mandatory standards forbid the employer and employees of a single firm from engaging in transactions that might depart from the norm mandated by regulations. For example, suppose that only two OSHA standards exist, a noise standard set at 90 dBA and a punch-press standard that specifies an automatic stock feed, and that the workers in a small punch-press shop are very experienced and very safety-conscious. They do not fear injury from the punch while their hands are in the die. They also hate noise. The two mandatory standards prevent workers from negotiating an agreement with management that would exempt management from compliance with the punch-press standard in return for a noise level lower than 90 dBA.[22] National standards can only reflect the intensity of preferences of labor and management at the national level To the extent that local preferences depart from aggregate preferences, the current standard-setting process will force outcomes that might be improved upon by local agreement.

Obviously, economic efficiency is not the only criterion that should be used to evaluate the present method of defining solutions to problems posed by hazards in the workplace. We are also concerned with whether the process encourages consensus among the parties and whether it defines effective solutions. Consensus is at least evidence that the standard-setting process has considered the intensity of the preferences of the parties in defining hazard-abating solutions, and consensus or agreement is politically desirable because it reflects peaceful reconciliation and accommodation of competing interests. Also, under the present regulatory structure, consensus is important because of its impact on compliance. If employers believe that OSHA standards are unreasonable, the compliance rate will fall and OSHA resources will be spent fighting legal battles over the validity of the agency's action.

Labor and management do participate in many stages of the current standard-setting process, but the present regulatory structure does little to encourage agreement between them over the steps that should be taken to abate hazards. As noted before, rarely if ever do consensus recommendations emerge from a Standards Advisory Committee. The public hearing process, conducted in the manner of a judicial trial, is inherently divisive. And final decisions on the content of standards are made by agency officials acting in what is essentially a judicial capacity: sifting and weighing conflicting evidence and choosing what seems to be a defensible solution. That the final standard does not reflect a meeting of the minds is best evinced by OSHA's litigation record. Eight of OSHA's first fourteen standards were challenged in the courts, in some cases by both labor and management.

The driving force behind some of management's challenges to OSHA regulations is a widely held belief that many of the standards are overly complex and unworkable. The complexity is endemic to a regulatory structure that relies upon standards and inspections to abate job hazards. Because vague standards that create large potential civil liability may be unconstitutional, and because nonspecific standards place excessive discretion in the hands of untrained OSHA inspectors, the present regulatory structure creates considerable pressure for OSHA to promulgate regulations that are very detailed and thus sometimes very confusing.

Complexity in standards would be tolerable if it yielded effective hazard-abatement solutions. Unfortunately, it is very difficult to prescribe general remedies for the abatement of workplace hazards that will work well in a broad range of specific situations.[23] Even where hazards are general and not workplace-specific, as are many occupational health problems, implementation of abatement measures is likely to vary considerably among firms. For example, OSHA has considered a standard that would mandate annual physical examinations for workers exposed to certain health hazards. When the exam indicated that a worker had received excessive exposure to the hazard, the employer would be required to transfer the worker to another job. In a large unionized firm with many different types of jobs, such a standard might be feasible. Presumably, the collective bargaining agreement would protect the worker's right to continued employment.[24] However, in a smaller nonunionized firm with few alternate jobs the standard would not work. Without job security, an overexposed worker might quickly find himself out on the streets, and to mandate job security in the standard might drive a small firm with few alternate jobs out of business.

The current process of solution design does attempt to integrate some implementation considerations into the standard-setting process; much of the testimony presented at the hearings is devoted to implementation problems. However, this is largely futile. Government regulations and standards are inherently blunt instruments. It is impossible to write them with enough specificity to reasonably accommodate all possible implementation scenarios without making them incomprehensible. Moreover, efforts to "fine tune" standards will delay further a process that already proceeds at a glacial pace.

In summary: The current method of defining hazard-abatement solutions through standard setting does not perform well by any criterion. The process does not efficiently indicate the resources to be committed to hazard abatement, nor does it allocate them optimally. Mandatory standards also prevent intrafirm shifts of hazard-abatement resources that might benefit both labor and management. The process is inherently divisive and produces standards that are often too complex to be understandable and yet too simple to be effective. Finally, the process is slow and cumbersome.

Implementation of Abatement Procedures
The present regulatory system relies on inspections and fines to induce employers to implement the abatement procedures defined in OSHA regulations. The threat of fines is expected to induce some firms to comply with health and safety regulations prior to inspection (OSHA disingenuously calls this "voluntary compliance"), and the threat of being fined as much as $1,000 per day for failure to abate a hazard after inspection is expected to bring violators into compliance (we will call this "coercive compliance"). In a sense, both types of compliance are coercive in that they induce employers to take actions they would not take in the absence of regulation. What distinguishes the two is the immediacy and the magnitude of the threat.

The incentives created by the present system for voluntary compliance are small. As noted in chapter 1, OSHA fines are modest and inspections infrequent. The average first-instance fine proposed in 1975 was only $144 per inspection. Less than 2 percent of all firms are inspected annually. By contrast, compliance costs are often substantial. The National Association of Manufacturers survey cited earlier estimated that firms with fewer than 100 employees would spend an average of $35,000 to comply with OSHA regulations. At any reasonable cost of capital, it will almost always be advantageous for firms to delay implementing OSHA-prescribed abatement procedures until forced to comply by inspection. It is difficult to make the threat of OSHA inspection so onerous that it will encourage widespread voluntary compliance. Although doubling the size of the inspection force would double the expected cost of noncompliance, that would still only raise it to $6 per year. And increasing the size of first-instance fines probably is both unjust and infeasible. To make the threat of inspection credible, first-instance fines would have to be made exorbitant. It could be called inequitable (and unconstitutional) to single out relatively few employers for inspection and then saddle them with ridiculously high penalties for running their businesses the same way everyone else does—in violation of OSHA regulations.[25] Moreover, Congress would have to approve any increase in the statutory limits on penalties, and this seems unlikely in light of congressional opposition to proposals to increase the size of the OSHA inspection force.

Recognizing the inherent limitations of voluntary compliance, OSHA has attempted to maximize the returns on coercive compliance by targeting inspections at high-risk workplaces. Even coercive compliance, however, has its limitations. With OSHA's inspection resources, fewer than 100,000 workplaces per year can be coerced into compliance. Even coercive compliance cannot ensure total implementation of the hazard-abatement procedures defined in OSHA regulations. Very few hazards can be eliminated completely by physical changes in the workplace alone. To be effective, physical changes must almost always be accompanied by parallel changes in work practices and routines. At a minimum, passive health and safety equipment must be adequately maintained. Some safety devices must not be consciously circumvented (antishock interlock switches, machine guards, etc.); others must be actively employed when demanded by the work situation (such as safety glasses, temporary railings to protect roofers, and hard hats). Other health and safety equipment requires constant monitoring of loads and exposure levels. Some abatement procedures require no physical changes in the work environment and can only be implemented by changing work routines (for example, limitations on the height of free-standing stacked material, and other housekeeping requirements).

OSHA inspectors observe the workplace at a single point in time. They can observe whether required capital equipment is in place and to some extent whether it is being used. It is difficult, however, to determine whether the equipment has been adequately maintained, whether it is always used properly, and whether conditions that exist on the day of inspection are representative or unique. The present regulatory structure has a very limited ability to induce the changes in work practices necessary to ensure effective implementation of hazard-abatement procedures. OSHA's presence in the workplace is limited to a maximum of one or two visits a year. During inspection, an OSHA inspector may order a temporary railing to be constructed or fine a firm for not having proper coke-oven emission control equipment. He cannot, however, order the resolution of a labor dispute arising out of a change in job assignments attributable to an OSHA regulation, or ensure that supervisors will vigorously enforce health and safety rules in the future. Continual and regular implementation of hazard-abatement procedures requires a system of continual and regular incentives, properly placed. The only continual incentives provided by OSHA result from the threat of reinspection. As noted before, these incentives are modest and directed exclusively at the general employer. The present regulatory approach does little to provide lower-level supervisors and workers with direct incentives to assist in implementing abatement procedures.[26] The only hazards likely to be eliminated under the current system are those that can be permanently abated by order of an inspector who visits the plant at most twice a year.

Conclusions

Much of the public criticism of OSHA has centered on the slowness of the standard-setting process, the infrequency of the inspections, and the size of the fines. The agency has been called stupid, poorly managed, and insensitive to the concerns of both labor and management. Much of this criticism is misdirected. OSHA has been relatively ineffective to date because the existing regulatory structure is simply not capable of addressing the millions of separate problems that constitute the occupational safety and health problem in the United States. Most safety hazards are not subject to control by standards. Uniform standards necessarily result in inefficient allocation of hazard-abatement resources. Regulations cannot be written with enough specificity to accommodate all of the unique conditions encountered in the nation's five million workplaces. Even conscientious and well-trained inspectors can only observe a small proportion of hazards. And fines do not create the necessary incentives to ensure that both workers and managers will properly implement procedures designed to guarantee the safety of the workplace. In short, occupational safety and health policy is ineffective because it is poorly designed.

4
Doing Better

Criteria for Effectiveness

Any policy that has as its goal the elimination of hazards from the workplace must have the following characteristics to be effective:
- It must be capable of affecting a broad range of dangers, including hazards created by acts of workers, temporary hazards, synergistic hazards, and hazards specific to individual workplaces.
- It must encourage for each hazard the development of an abatement strategy that is cost-effective, reflects the preferences of the workers it is designed to protect, and respects the unique implementation problems encountered in diverse workplaces.
- It must provide adequate incentives for employers to undertake the necessary investments.
- It should provide incentives for workers and supervisors to alter their conduct as required.
- It should adapt quickly to technical and economic changes that may necessitate further abatement procedures or reduce the efficacy of those already in place.

Alternate Strategies

Because OSHA has been the subject of such intense criticism in the academic and the popular press, there is no shortage of suggestions for how policy might be improved. Respected commentators have proposed abolishing the agency entirely,[1] substituting an injury or exposure tax,[2] refining the hazard-abatement incentives offered by workmen's compensation,[3] streamlining the standard-setting process,[4] improving the targeting of inspections,[5] and encouraging hazard abatement by stimulating labor and management to address health and safety issues through collective bargaining.[6] This section reviews each of these approaches in light of the operational objectives defined above and in view of the political considerations discussed in chapter 2.

Abolition of OSHA

It is difficult to take this suggestion seriously from a political perspective. Although OSHA has its opponents, Congress and the public seem far from willing to abandon regulation of job hazards entirely. Nonetheless, it is worth pausing to evaluate this suggestion, if only because it gives us some sense of how well-functioning private markets might operate to achieve health and safety objectives.

In a world where workers were fully informed of occupational risks, wage premiums would be paid for hazardous work. By demanding extra pay for dangerous jobs, workers themselves would signal the existence of different types of job hazards to employers. In theory these costs as well as the costs of accidents (delays, damaged equipment, compensatory payments to workers, etc.) should motivate employers to take steps on their own to abate hazards. If markets worked perfectly, the self-interested employer would respond to job hazards in a cost-effective manner that would consider both the preferences of his workers and his own costs of abatement.[7] Since costs of implementation are relevant to the employer, they would also be considered in the solution design. Perhaps the most desirable feature of perfect markets is that they create real incentives for employers to continually and effectively implement hazard-abatement solutions. If an employer failed to implement effectively a particular hazard-limiting program, his workers would again demand compensating wage differentials. In effect, the employees would act as a constant inspection force. If markets worked perfectly (and instantaneously), the employer would incur costs immediately for failure to implement an abatement procedure, and this would be the incentive for continual attention to implementation and to any changes in working conditions that might create new hazards.

Unfortunately for all concerned, markets work neither perfectly nor instantaneously. Workers rarely have the information at hand to intelligently assess risks. Because wage rates are not fluid, workers cannot easily command hazard pay and employers cannot recognize wage-bill reductions due to hazard abatement. And since health care is subsidized to some extent by government, employers and employees do not bear all of the costs of occupational accidents and illnesses; thus, their private decisions will not necessarily lead to socially optimal reductions in job hazards.

These difficulties notwithstanding, opportunities do exist for achieving some job safety and health policy objectives through efforts to improve the working of market mechanisms. It is difficult to loosen the institutional rigidity of wages, but government could

Doing Better

do a better job of ensuring that workers are informed of risks, as a number of analysts have suggested.[8] Better information should at least encourage workers to distinguish among employment opportunities on the basis of risk, and this would create some additional pressure for management to abate hazards. From a policy perspective, providing more information is cheap, politically unobjectionable, and a useful supplement to just about any regulatory program. However, because information provision only addresses one way in which markets fail, it is not a substitute for a government incentive program or a direct system of regulation.

Regulation through Injury Taxes

One way of overcoming the dulling of incentives by wage rigidity is to tax undesirable health and safety outcomes—specifically, job accidents and high levels of exposure to health-threatening substances.[9] In theory, such a tax should motivate employers to identify hazards and adopt cost-effective abatement procedures. Because the tax would be levied on outcomes, employers would have an incentive to ferret out all sorts of different hazards. Furthermore, such a strategy would free the employer to respond to hazards in a manner both cheap and consistent with implementation considerations.

There are a number of flaws in this approach. To tax effectively requires an ability to monitor taxable events with accuracy. Accidents are easily observed and relatively well reported, but constant monitoring of exposure levels is difficult and expensive. The equipment is costly and not uniform, and different tests must be used to detect different substances. NIOSH has estimated that the annual cost of monitoring exposure to toxic substances ranges from $675 million to $2 billion.[10] Moreover, the cost of detection is insensitive to the size of the firm or the level of exposure. Consequently, an exposure tax might impose substantial measurement costs on small firms, even if their actual hazards were minimal.

Even taxing accidents presents difficulties. Because accidents are stochastic events, an accident tax would only impose expected future costs on employers who failed to abate hazards. As experience with similarly stochastic OSHA fines suggests, an accident tax would have to be extremely large to induce employers to invest in expensive abatement technology. Furthermore, it is not clear that employers would respond to an accident tax even if the expected amount of the tax were larger than the cost of compliance. The basic assumption underlying all incentive taxes is that the institution that pays the tax will be sensitive to its imposition. In a frictionless, comparatively static world this assumption might be reasonable, but the world is neither frictionless nor static. Institutions tend to be organized to perform the tasks they are currently performing; their capacity to perform new tasks is limited. Moreover, they learn slowly and can only pay attention to a few things at a time. Unless an organization were structured to recognize and respond to changes in its organizational environment, the marginal change in costs brought about by the imposition of an accident tax would probably go unnoticed.[11] This would be especially true if the costs attributable to the tax were buried among other changes in operating expenses.

In short, taxes are ineffective if organizations lack the institutional equivalent of sensory receptors. Because accidents are stochastic events, an accident tax would only impose expected future costs on employers who failed to abate hazards. Unlike an exposure tax or the costs imposed by instantaneously adjusting wage differentials, an accident tax would not impose immediate costs on an employer who failed to implement abatement procedures. The employer would pay only after the untoward event had occurred. Unless the tax were very large, it would not create sufficient incentives for continual attention to implementation. Finally, an accident or exposure tax is not politically feasible as an alternative to OSHA-type intervention. The public does not understand how such taxes work. The taxes would not appear to guarantee any particular type of action on the part of management. And labor opposes such taxes on the grounds that they would give employers a "license to maim."

Regulation through Workmen's Compensation
Workmen's-compensation laws enacted by each of the states make employers strictly liable for injuries or illnesses incurred by their workers in the course of employment.[12] Claims are processed through an administrative procedure that typically provides a fixed

schedule of awards for specific types of disability. Employers are required to carry insurance to cover the costs of claims.[13] The insurance premiums are determined in part on the basis of the risks present in the insured firm as well as its prior claims experience. Banks pay lower rates than foundries, and to some extent foundries with good accident records pay less than foundries with bad accident records. Because premium levels are determined in part by the level of risk in the insured workplace, workmen's compensation insurance provides financial incentives for hazard abatement similar to those that would be provided by an injury tax. By reducing hazards (and the likelihood of a workmen's-compensation claim), an employer can reduce his insurance premiums.

Though the rating process could be refined to sharpen the hazard-abatement incentives exerted by workmen's compensation, such a system would still not achieve many of the objectives of a good job safety and health policy. Because workmen's compensation operates on the same analytic principle as an injury or exposure tax, it suffers from exactly the same drawbacks: Risks are hard to monitor, incentives created are remote, and it is not clear that employers would respond to small changes in the cost of insurance premiums.

Furthermore, there are problems peculiar to regulation by workmen's compensation. From an analytic perspective, a tradeoff exists between the efficiency gains realized from spreading risks through insurance and the safety incentives created by permitting premiums to fluctuate as a function of past experience.[14] An optimal package of experience rating and coinsurance would not offer as sharp an incentive for hazard abatement as an injury and exposure tax. Also, relying on workmen's compensation to create hazard-abatement incentives has even less potential popular appeal than an injury tax, because it has a lower political profile. Furthermore, labor perceives workmen's compensation as a compensatory program, not as an incentive system to make the workplace safe. There are small gains to be had from improved rating of workmen's-compensation premiums, but even if these improvements were to be coupled with the existing regulatory program we would still be a long way from achieving the objectives of an effective job safety and health policy.

Improving Standard Setting and Inspection

OSHA's standard setting and enforcement activities could be improved in lots of different ways. More attention to implementation problems, better consideration of costs and benefits, and better targeting of inspections at high-risk workplaces would all produce improvements in OSHA's performance. However, such efforts, even if successful, would not address the structural deficiencies in the existing regulatory approach. Even if all standards were cost-effective a regulator would only be able to influence a small proportion of all occupational hazards. Though better targeting of enforcement resources might increase the effectiveness of coercive compliance, it would not create the continual pressure that is needed if standards are to be implemented effectively. Even if OSHA's resources were increased substantially, the agency still would not be able to address most workplace-specific hazards. OSHA's primary problems are structural, and although marginal improvements in the existing program would yield welcome benefits they would not resolve the major structural problems.

Regulation through Collective Bargaining

Much of what a labor union does is very similar to what a regulatory agency such as OSHA does. Both institutions exist to alter the conduct of management. Both are primarily concerned with improving the lot of workers. Both spell out in great detail the dimensions of acceptable management conduct, in one case through regulations and in the other through the terms of the collective bargaining agreement. And both unions and OSHA attempt to ensure management compliance through inherently coercive processes.

There are also a number of significant differences between union activity and regulatory activity. First, a union exists almost exclusively to represent and further the interests of its members. Contact between officers and members is frequent, especially at the local level. If the union ignores the needs of the rank and file, the leadership may be voted out of office. OSHA, on the other hand, represents a large and diffuse class of people who have little if any contact with the agency. It is insulated from its constituency—workers—by size and distance. Second, unlike OSHA, a union cannot unilaterally mandate changes in working conditions. Each change must first be negotiated with management. As a result, the rules governing the conditions of employment produced by this bargaining represent a reconciliation of competing interests by the principals. In contrast, OSHA functions more like a judge, sifting and weighing evidence and rendering decisions that often satisfy nobody. Third, unions deal with a much broader range of issues than OSHA, including wages, hiring, layoffs, seniority, insurance, training, education, promotion, incentives, and discipline, in addition to some health and safety issues. Related problems can be solved in a single forum. Fourth, the work rules developed through labor-management negotiation do a better job of accommodating the diverse conditions encountered in different workplaces than uniform national standards. There are approximately 150,000 separate collective bargaining agreements in force in the United States. A majority of organized workers work under contracts negotiated with a single employer or for a single plant.[15] Fifth, labor-management rulemaking is more flexible than government regulatory activity. Collective bargaining agreements are renegotiated periodically, and the parties may modify rules during the term of the contract by mutual agreement. Sixth, the coercive enforcement processes employed by unions and OSHA differ substantially across a number of dimensions. Collective bargaining agreements are enforced by the workers they are designed to protect, whereas regulations are enforced by relatively disinterested inspectors. The union's presence in the workplace is continual and pervasive; OSHA's is severely limited. Unions can vary and escalate their threats and sanctions; OSHA cannot. As a participant in the operation of the workplace, a union can influence events that affect the cost of implementing changes in working conditions—something that OSHA is powerless to do. Union demands are treated relatively seriously by foremen and supervisors; OSHA regulations and fines are not.

The characteristics that distinguish collective bargaining activity from regulatory activity suggest that unions may be a convenient vehicle for achieving job safety and health objectives. Collective bargaining agreements produce flexible rules governing working conditions that are tailored to meet the needs of the individual firm, that reflect the preferences of the parties, that consider the costs of implementation, and that are cost-effective. More important is the fact that the rules are enforced.

Consequently, from a policy perspective, it may make sense to involve unions in abating hazards in the workplace, rather than relying exclusively on governmental regulation and inspection. There may be gains to be realized from encouraging unions to pursue goals consonant with federal policy. To the extent that labor and management can be induced to negotiate health and safety rules within the context of the collective bargaining agreement, we can decentralize some aspects of the regulatory intervention mechanisms almost to the plant level. In theory, this should produce investments in occupational health and safety that are both more efficient and more effective than those produced by the present system alone.

Conclusions

Of the regulatory alternatives discussed, only collective bargaining appears capable of addressing the structural deficiencies that have plagued OSHA to date. Yet even if the federal government were to be successful in encouraging labor and management to be more active in pursuing health and safety objectives through the collective bargaining process not all of OSHA's problems would be solved. Only 28 percent of American workers belong to labor unions. Hence, even this strategy has limited potential.

The one clear lesson that emerges from this chapter is that no single regulatory approach is likely to be successful. Instead, we may need to adopt a mixed strategy that takes advantage of a number of opportunities to marginally improve the performance of the federal regulatory effort. At the same time, it is probably wise to lower our expectations of what any program for regulating occupational safety and health is likely to accomplish.

III
BARGAINING AS A REGULATORY STRATEGY

5
From Theory to Practice

In the preceding chapter we suggested that collective bargaining offers a number of structural advantages over "command and control" regulation in limiting workplace hazards. So far, however, the discussion has been largely abstract. It has not addressed practical questions such as which regulatory functions might actually be delegated to the collective bargaining process, whether increasing the involvement of unions in regulating job safety and health will improve the lot of workers, or whether there exist practical barriers that discourage labor unions from taking a more active role in eliminating hazards from the workplace. These questions are of very real concern—a major reason why Congress passed the Occupational Safety and Health Act was because it did not believe that market forces, including collective bargaining, were capable of adequately controlling job-related accidents and illnesses.

The first section of this chapter describes the direction of bargaining over health and safety issues in three different unions. These cases illustrate the types of changes that might be expected from successful collective bargaining activity. The second section reviews aggregate measures to determine the scope of bargaining over health and safety issues in other industries. The last section summarizes which health and safety functions labor and management appear capable of performing jointly, and which are beyond the scope of the collective bargaining process.

The best way to see whether joint labor-management health and safety efforts improve the lot of workers is to analyze accident and illness frequency statistics. If the theory is correct, firms with negotiated health and safety agreements should have lower frequency rates (all other things being equal) than firms without such agreements. It should also be possible to study the incidence of occupational disease and accidents over time as new hazard-limiting agreements are negotiated. Neither approach is attempted in this book. To draw meaningful inferences from either type of analysis requires access to sensitive data that cannot be obtained without the cooperation of many separate firms. Even firms with very good safety records are reluctant to disclose detailed accident and exposure statistics.[1]

Because of the constraints imposed by the unavailability of data, we will focus on three case studies of how collective bargaining affects intrafirm hazard-abatement processes. The cases have been selected for their diversity. The United Auto Workers–General Motors agreement is interesting because it provides for the training of full-time union health and safety stewards at the local level. The United Steelworkers and U.S. Steel, on the other hand, have concentrated on negotiating detailed local agreements that provide for the abatement of specific hazards. Unlike the UAW, the steelworkers also arbitrate many health and safety grievances through an expedited arbitration process. The United Association of Plumbers and Pipefitters was selected because, as a craft union dealing with hundreds of employers, it has concentrated its health and safety efforts on training and educating workers.

The United Auto Workers and General Motors

General Motors, the second largest corporation in the United States, employs close to 800,000 workers in more than 120 plants. It does substantially more than assemble motor vehicles; it also produces many of the parts that go into them, including batteries, air conditioners, headlights, radios, and cigarette lighters. Approximately 420,000 GM workers are represented by the United Automobile Workers in 143 separate bargaining units. UAW members work in stamping plants, foundries, chemical plants, research centers, and assembly plants. In many respects the UAW workers are a microcosm of the American manufacturing workforce.

In 1973, as part of their national contract negotiations, the UAW and GM negotiated a unique memorandum of understanding on health and safety that provided for the creation of full-time union health and safety representatives in GM plants with more than 600 employees.[2] The representatives are selected by the international union from a list of names submitted by the local. Many of these individuals have gained experience from participation in health and safety committees. Initially, they receive about 40 hours of training in hazard recognition at the General Motors Institute; the corporation's own health and safety representatives receive the same training. The curriculum was established jointly by GM and UAW industrial hygienists, who also serve as instructors. Union representatives are taught how to read a noise meter, how to operate a smoke detector, how to take air-quality samples for testing elsewhere, and how to recognize occupational hazards commonly found in their respective manufacturing environments. OSHA complaint procedures and the details of relevant OSHA regulations are also part of the curriculum. Many UAW health and safety representatives supplement this training by taking related courses at local universities and participating in professional seminars. General Motors pays the training expenses and the wages of union health and safety representatives and supplies the necessary testing equipment.

The memorandum of understanding describes the duties of the union representative only generally. With his management counterpart (usually the plant safety director), he inspects the workplace once every two weeks, accompanies federal and state OSHA inspectors on inspection tours, reviews accident reports, advises employees about exposure to hazardous substances, and reviews and recommends safety education and training programs for the workers. A provision added during the 1976 contract negotiations empowers the union representative, acting jointly with the plant safety officer, to order the shutdown of any equipment or operation that poses imminent danger to employees.

In addition to the duties described above, the union representative also has an important role in resolving disputes over health and safety issues. The memorandum of understanding created a special procedure for processing health and safety complaints. To comprehend this new process, one must understand the traditional procedure used to resolve contract grievances.

The traditional GM-UAW grievance procedure is divided into four different steps, and lower-level decisions are reviewed by higher authorities at each successive step. At the first step, an employee states a grievance verbally to his foreman. The employee may be accompanied by his committeeman, an elected officer who has specific knowledge of the collective bargaining agreement.[3] If the grievance is not settled verbally, the committeeman writes it up and takes it to a higher supervisor (usually the general foreman). If the grievance is not settled by the general foreman, it may progress to the second step. At the second step the grievance is considered and investigated by the shop committee, an elected union body of between three and eleven members.[4] If the grievance is deemed meritorious, it is taken up at a meeting between the shop committee and a representative of the highest local management (typically the plant personnel director), who must render a decision in writing within fifteen working days. If the grievance is not settled at the local level it may move to the third step, which involves for the first time representatives of the international union and the corporation. Briefs are prepared by the local union and local management setting forth their positions. The international union may investigate the grievance to determine if it is meritorious, and if it is found meritorious it may be appealed to division management. If the grievance cannot be settled at this level, it may be appealed to an impartial umpire as the fourth step. Most grievances are settled short of this fourth step; in 1976 only 29 out of 241,000 grievances filed by UAW members went to arbitration.[5]

The memorandum of understanding altered the processing of health and safety complaints in a number of important ways. First, it gave union committeemen the power to call in the union health and safety representative to deal with the foreman at the first step. Because the representative knows the OSHA regulations and can take relevant measurements, his presence improves the bargaining position of the committeeman relative to the foreman. Perhaps more important, it also improves the committeeman's authority to turn away groundless complaints. Committeemen, as elected officials, often feel compelled to respond to the demands of their constituency by writing up a grievance, and find it hard to inform a member of the rank and file that his beef with management has no merit. In contrast, the health and safety representative is appointed by the international union and is barred from participation in union politics;[6] thus, he can turn off groundless complaints while allowing the committeeman to give the appearance of vigorously pursuing his constituent's interests.

Second, the memorandum of understanding empowers the union representative and his management counterpart to settle jointly complaints that are not resolved at the first step. Since the two can also order hazardous operations shut down, this provision creates incentives for foremen to settle health and safety disputes before they reach the second step. It also puts individuals who are exclusively concerned with improving health and safety in the workplace in a position to decide disputes. This gives health and safety a higher priority on the agenda of both labor and management than it would have if such decisions were made by the shop committee and the plant personnel director alone. However, the chairman of the shop committee and a representative of higher management are brought in if the union representative and his management counterpart are unable to resolve the dispute. Health and safety complaints that are still not settled at this stage are referred directly to the third step of the grievance procedure, which involves representatives of the international union and the corporation. Health and safety complaints reach this third step in a maximum of two or three weeks.[7] Because health and safety issues are not arbitrable under the GM-UAW national agreement, this is the last step in the complaint procedure. If the two sides fail to reach agreement the union is free to strike during the term of the contract, notwithstanding the no-strike pledge. Since the international union must issue a notice of intent to strike five days before commencing the walkout, the international union can effectively veto a local strike if it determines that it is motivated by illegitimate health and safety complaints.

The third effect of the memorandum of understanding is to give union representatives the power to initiate complaints when conditions adversely affect substantial groups of employees. This is an important power when coupled with the union representative's inspection and dispute-resolution functions. It means that the representative and the plant safety officer can inspect the workplace, identify hazards, negotiate between themselves over abatement procedures, and order and enforce implementation of solutions.

In practice, the influence a union health and safety representative exercises within the plant is determined by his personal relationship with his management counterpart. Nearly all of their powers are shared. Inspections are conducted jointly, training programs are planned jointly, the decision to shut down an operation must be made jointly, and disputes are settled jointly. As a result, much of a union representative's substantive power depends on his ability to persuade the plant safety officer of the merits of his arguments. Unlike the union representative, who is almost independent of his constituents, the plant safety officer is very much an agent of management. He is evaluated and promoted according to management's appraisal of his performance. While his primary concern is health and safety, he still must be sensitive to cost and efficiency.

A union representative can influence a plant safety officer's judgment on health and safety matters by appealing to the safety officer's professional biases or to his friendship, by initiating (or failing to turn off) a number of groundless health and safety complaints, by threatening to file a complaint with OSHA that would trigger an OSHA inspection and possible fines, or by attempting to get local and international support for a possible strike. The first approach only makes the plant safety officer feel bad, the second makes him look bad to management, and the rest impose substantial costs on management (if the threats are credible). Because a strike also imposes costs on the union, it is rarely used to obtain health and safety concessions. Only five notices of intent to strike have been issued by the international union exclusively over health and safety issues since the first memorandum of understanding was negotiated in 1973, and all of the disputes were settled short of a strike.[8] As a result, the threat of complaining to OSHA is often the easiest way for a union representative to put substantial pressure on his management counterpart.

The effectiveness of the plant safety director and the union representative can vary according to how well the two work together. If they can solve problems efficiently and fairly, they may find themselves involved in a broader range of issues than is specified in the memorandum of understanding. In some plants the union and management representatives are consulted on virtually every dispute with health and safety dimensions, and in some cases they are the ones who negotiate the health and safety aspects of the local bargaining agreement.

General Motors originally opposed the concept of a company-funded union health and safety representative as an encroachment on corporate authority. However, company officials at a number of different levels have expressed satisfaction with the program. International and local union officials are also pleased. In some respects, the nearly universal approval of this program should not be surprising. It does a little something for everyone at a relatively modest cost.[9] From GM's perspective, the foreman benefits by having to handle fewer bogus complaints. In addition, the union representative sometimes proves helpful in getting recalcitrant workers to comply with company health and safety rules. To the extent that the union representation program has elevated health and safety on management's agenda of considerations, it has improved the status of the plant safety director. To local management, the union representative helps to ensure that supervisors implement health and safety procedures as directed by management. One personnel director noted that he preferred to deal with the union on health and safety matters rather than with OSHA, which he viewed as irrational and unpredictable. Although GM will not release any statistics, it acknowledges that the negotiated health and safety program has improved conditions. This view is shared by the union.

The union also benefits in a number of ways from the health and safety program. Clearly, to the extent that hazards are eliminated, the rank and file are better off. The health and safety representative gets a lifetime job, superseniority, an office in the plant, and quasi-professional status. The union representative's role allows the district committeeman to be responsive to the needs of his constituency without having to promote false grievances. The union local gets one additional company-funded job to use as patronage, as well as more effective bargaining and dispute settlement with regard to health and safety issues. The benefits to the international union are mainly derivative. About the only individuals within the UAW who have expressed dissatisfaction with the negotiated health and safety program are a few members of shop committees who have lost some influence, since the special health and safety compliance procedure allows the union representative and the chairman of the shop committee to settle grievances.

In summary: The GM-UAW memorandum of understanding has brought about five changes in the structure of local bargaining over health and safety issues. It has created a health and safety lobby within labor and management at the plant level. It has given the union the capacity to assess information independently and bargain intelligently over job hazards at the local level. It has expedited the resolution of health and safety disputes. By involving health and safety personnel in dispute resolution it has encouraged resolution of such disputes according to the merits. And because the memorandum forces labor and management to share responsibilities for administration of the health and safety program it has fostered a cooperative attitude between the union and the company on health and safety matters.

The United Steelworkers and U.S. Steel: The Clairton Agreement

Health and safety bargaining in the basic steel industry contrasts sharply with the example of the auto workers and GM. Health and safety disputes are "strikable" issues during the term of a contract in the auto industry, but under steel-industry contracts such disputes must be arbitrated. While the UAW has focused on establishing health and safety stewards as a means of controlling job hazards, the United Steelworkers of America have negotiated detailed agreements that provide for engineering controls and job rotation. The UAW has concentrated on identifying hazards plant by plant; the Steelworkers have relied on industrywide studies of occupational hazards. Both the nature of the hazards and the structure of collective bargaining in the steel industry account for the difference in approach between the United Steelworkers and the United Auto Workers.

The dirtiest and most hazardous jobs in basic steel are probably in coke-oven plants. Coke is the cellular residue that results from distillation or carbonization of coal. It is used primarily as a fuel and reducing agent in blast furnaces. Coke is produced by heating coal to very high temperatures (1,900–2,000 degrees F) in batteries of tall narrow ovens. Coal is introduced into the oven from above by means of a "larry car." The lid of the oven being charged is removed and the larry car deposits powdered coal into the oven. After the lid is replaced, the oven is fired for 16–20 hours until the coal is "coked." After coking, the doors on the front and the back of the oven are removed and the coke is pushed out of the oven into a quench car by a ram. The quench car is rolled under a quench tower, where the hot coke is cooled with water. The cooled coke is then removed to the blast furnace.

Each of the three major operations involved in manufacturing coke produces harmful emissions. As the oven is charged with powdered coal, coal dust and oven gases inevitably escape into the working environment. More noxious vapors are produced as the coke is pushed and quenched. Workers exposed to these emissions have an increased risk of developing lung cancer and urinary-tract cancer. The magnitude of the excess risk depends on both the duration and the location of exposure. For example, a topside coke-oven worker with more than 10 years of experience is 10 times as likely to contract lung cancer as a steelworker of similar age and experience working in a noncoke environment.[10]

The largest coke-oven plant in the world is located in Clairton, Pennsylvania. It is operated by U.S. Steel, and its workers are organized by Local 1557 of the United Steelworkers of America. It is nearly twice the size of the next largest coke plant in the United States. In 1974 the United Steelworkers and U.S. Steel concluded negotiations on a detailed local agreement designed to reduce worker exposure to coke-oven emissions at the Clairton plant. The foundation for the agreement was laid during the 1968 contract negotiations, when the Steelworkers successfully bargained for in-plant air-quality testing. The tests confirmed the suspicions of many coke-plant workers: Worker exposure to coal tar pitch volatiles exceeded recommended levels, in some cases by a factor of 20. After passage of the Occupational Safety and Health Act the Steelworkers filed suit to force OSHA to enforce the Threshold Limit Value standard for coke-oven emissions at the Clairton plant. Using this suit as leverage, in 1974 the Steelworkers got U.S. Steel to agree to negotiate a program to reduce worker exposure to coke-oven emissions. In return, I. W. Abel, the international president, agreed that the union would relinquish party status in the OSHA suit. The agreement also stipulated that if the parties were unable to reach agreement on a coke-emissions program unresolved issues would be settled through arbitration. This provision proved unnecessary. Negotiators for the international union and U.S. Steel reached agreement on November 6, 1974.

From Theory to Practice

The first section of the four-part Clairton Memorandum of Agreement deals with job rotation. To reduce exposure to harmful emissions the agreement gives each member of the coke battery crew an additional 30 minutes of "general spell time" (break time) per shift. To cover for workers on spell time, the agreement authorizes the hiring of 2.5 new utilitymen per crew. Because the Administrative Controls section increases the duties of the door cleaners, the agreement also authorizes the hiring of one additional door cleaner per crew. The net effect of these changes is to increase the size of coke battery crews from 17.5 to 21 workers. The changes in crew size called for by the agreement were viewed as union victories for two reasons: The reductions in work time make the lives of coke-oven workers a little bit easier, and the increases in crew size create new job opportunities for union members (even though the jobs are in dirty coke-oven plants).

The union could not have "sold" the Clairton agreement to the rank and file without first adjusting the formula used to determine incentive pay. Coke-oven workers are paid according to the size of the work crew and the number of ovens pushed per shift. Because of the expanded door- and lid-cleaning responsibilities mandated by the agreement, the number of ovens pushed per shift dropped from 90 to 83 despite the 20 percent increase in crew size. Thus, part 2 of the agreement adjusts the incentive pay formula so the coke-oven workers do not have to take a pay cut due to the decline in their productivity precipitated by the administrative and engineering controls.

Part 3 of the agreement describes engineering improvements in the design of the ovens intended to limit emissions. A coke oven is basically a large box with three sides removable to allow the coal to be dropped into the oven from above and the coke to be pushed out. The level of emissions is determined by the tightness of the doors and lids, the design of the larry car, and the efficiency of the aspiration system that draws off by-product gases. As part of the agreement, the company pledged to improve the mechanical design of each of the above systems and to test the feasibility of filtering the air in the larry-car cabs and in the lunch areas.

Mechanical improvements in the door and lid seals cannot by themselves ensure a perfectly tight oven. Carbon buildups that occur during the coking process also inhibit tight metal-to-metal seals. Consequently, maintenance plays a critical role in emissions control. Part 4 of the agreement specifies detailed administrative and operating practices and controls intended to provide proper emissions-related maintenance and practice. Before each charge the doors and lids are to be inspected and cleaned, tar and carbon deposits are to be removed from the aspiration system, and the larry car is to be checked to ensure that the coal is properly distributed among its four hoppers.[11]

The last part of the Clairton agreement gives the employees the right to obtain copies of the records of their annual physical exams and x-rays administered by the plant physician. Surprisingly, this disclosure stipulation was an industry first.

Despite its breadth, the Clairton agreement ignores a number of significant issues. For example, there is no mechanism for resolving disputes arising under the agreement. The union did not push for a resolution procedure at the time of negotiation because it assumed that the agreement was simply an addendum to the regular Clairton local agreement,[12] under which disputes are arbitrable according to the procedures defined in the Basic Steel Contract. The union assumed that it would be allowed to take the company to arbitration if the company did not implement the agreement correctly. But the company's position is that the Clairton agreement represents an ongoing attempt at joint problem solving on the part of the company and the union, not a contract, and therefore disputes must be settled by the parties without arbitration. In practice, all disputes to date over implementation have been resolved through informal negotiations; thus, the parties have not yet been forced to resolve the question of arbitrability of the Clairton agreement.[13]

The Clairton agreement also does not include a performance standard for air quality in the plant. For example, the union could have demanded a company pledge to meet the OSHA-mandated threshold limit value for coal tar pitch volatiles in coke-oven plants; this would have provided insurance for the workers in the event that other controls proved ineffective. There are several reasons why the union did not push for such a performance standard. They viewed it as redundant in light of the company's legal obligations to comply with the OSHA threshold limit value. More important, it would have been extremely difficult for the Clairton local to enforce such a provision. Experience with the OSHA threshold limit value suggested that the company could easily come up with plausible explanations for failure to comply with that standard. Consequently, the union felt compelled to hold the company to specific engineering controls and changes in work practices. In addition, the union lacked the technical capacity to monitor compliance with a performance standard—a problem not posed by readily observable work practices and engineering controls. The union did ask for the right to monitor air quality, but was rebuffed by the company on the grounds that the Basic Steel Agreement already entitled the union to copies of periodic company air-quality studies.[14]

In view of the union's limited technical capabilities, it may seem surprising that the union did not ask for a UAW-type agreement that would have provided both training and hazard-monitoring functions for union representatives. There appear to be two reasons why the union did not ask for such a provision. First, the political structure of the United Steelworkers does not lend itself to agreements that appear to compromise the independence of the union. In contrast with the case of the UAW, all international and regional officers of the United Steelworkers are chosen in general elections by the rank and file. (The UAW officers are chosen by councils of locally elected officials.) Officials of the UAW are somewhat insulated from charges that they are too close to the company, whereas United Steelworkers officials are not. In fact, among the Steelworkers such charges are quite commonly leveled by challengers against incumbents. Thus, an agreement that provided for company funding and training of union health and safety officials might have been perceived as a company buyoff of the union. Second, as one union official reported, the local was reluctant to assume responsibility for enforcement of the health and safety agreement for fear of incurring liability in the event that it failed in its duties. In recent years, a number of other United Steelworkers locals have been sued by members claiming to have been injured because the union failed to discover a hazard in the course of an inspection mandated by the collective bargaining agreement.[15] None of these suits has been successful, but the threat of future suits has discouraged locals from negotiating agreements that give them substantial duties for health and safety inspection. Since under common law a union cannot incur liability for failure to detect hazards unless it has explicitly assumed such a duty, the union can avoid the threat of liability entirely by not negotiating an inspection clause.

The analysis of the UAW agreement at the beginning of this chapter credits much of that agreement's success to the fact that it places responsibility for enforcement in the hands of the most health-and-safety-conscious individual in the plant, the union health and safety representative. The Clairton agreement assigns the plant grievance committeeman to police compliance with the engineering and administrative controls specified in the agreement. Unlike the UAW representative, the Clairton grievance committeeman is elected, has no special training, and also holds a regular job within the plant. Health and safety matters must compete for his attention with other contract disputes, and this competition affects the likelihood of implementation of the work-practices provision of the Clairton agreement. As noted before, the level of emissions is determined in large part by the cleanliness of the doors, the lids, and the aspiration equipment. To ensure that workers have sufficient time to perform the cleaning responsibilities specified in the agreement, the grievance committeeman must keep the foreman from tightening production schedules too much. He must also press workers to perform unpleasant cleaning chores once they have been allotted the time by the foreman. The willingness of the grievance committeeman to push both the foreman and the workers to comply with the work-practices sections of the agreement is determined by the relative weight he gives to health and safety in comparison with other contract issues. Because the grievance committeeman is an elected union official, his priorities probably reflect those of his constituency, the rank and file.

Given the massive health and safety problems at the Clairton Works, one might expect to see a groundswell of rank-and-file support for vigorous enforcement of the relevant parts of the local agreement. In fact, Local 1557 is divided on the issue of the priority to be given health and safety on the union's bargaining agenda. While virtually all workers recognize the dangers associated with coke-oven employment, some believe that these dangers are inherent and cannot be eliminated through engineering controls and better work practices. This group would prefer that the union direct its efforts at winning higher wages and shorter hours, rather than expending scarce bargaining capital to achieve nothing but unpleasant work practices (door and lid cleaning) and expensive engineering controls. Union officials (unwilling to be quoted) estimated this group to comprise about one-third of the total membership, mainly older workers who would reap fewer benefits from improvements in health conditions. The difference of opinion within the rank and file may explain the local's reluctance to negotiate the Clairton agreement in the first place.[16] It also probably explains why there has been less than perfect compliance with the agreement on the part of both the company and the union (implementation of engineering controls has taken much longer than anticipated, and the union will admit privately that it is difficult to get workers to clean the doors and lids properly before each charge).

From a policy perspective, the interesting analytic question is not whether there has been perfect compliance, but whether the agreement has improved health and safety conditions within the plant. The answer to this question is a qualified "yes." Time-series data on in-plant emissions would be required for accurate assessment of the agreement's impact, and efforts to obtain this information from the company have been unsuccessful. Nonetheless, the agreement can be said to have resulted in many changes that should improve working conditions within the plant. For example, stage charging is now standard practice, maintenance of oven doors and lids has improved, more lid and oven cleaning is being done than before, there is one air-conditioned larry car and two air-conditioned hot cars, there are seven filtered-air lunchrooms, and there is one filtered-air rest area for lid men.[17] In addition, the negotiated increases in spell time have reduced individual exposure to coke-oven emissions.

In spite of these ostensibly desirable changes, the Clairton agreement may not have benefited Local 1557 collectively. By increasing crew size from 17.5 to 21 men per shift, the agreement has exposed more workers to the hazards of coke-oven emissions. Thus, unless the agreement reduces emissions enough to reduce each worker's probability of contracting disease by at least 20 percent the agreement will have increased the expected number of cases of disease.[18] Without precise knowledge of both the level of reduction in emissions attributable to the agreement and the dose-response relationship of the relevant diseases, it is not possible to assess the ultimate impact of the Clairton agreement on worker health. Nonetheless, the Clairton agreement has resulted in observable changes in working conditions within the plant—changes that are mostly favored by the rank and file.

Arbitration of Health and Safety Disputes in Basic Steel

With the exception of disputes arising under the Clairton agreement, all other health and safety disputes involving interpretation of contract language in the basic steel industry are arbitrable. For example, if a foreman asks an employee to perform a task that the employee believes to be dangerous, the employee has two options: He can perform the task and file a grievance or he can refuse to perform the task, ask to be transferred to another job, and then file a grievance. Since the company is not obligated to transfer a worker to another job, a worker confronted with an unusually dangerous assignment must make a difficult choice: He can perform the task and expose himself to possible injury or he can refuse the work and risk losing wages should the arbitrator rule against him.[19] Because of this incentive structure, if the right to refuse hazardous work is to be meaningful the arbitration process must be swift.

The Basic Steel Agreement provides an expedited process for arbitration of health and safety grievances. Health and safety grievances are initiated at the third level of the grievance process. This has two consequences: Two weeks of processing time are skipped; and since the "third step" involves review of the grievance by the general plant superintendent, the dispute is brought to the immediate attention of top plant management. A special arbitration procedure is employed for resolving health and safety grievances not settled by the parties. Pre- and post-hearing briefs are dispensed with, and formal rules of evidence do not apply during the arbitration hearing. Moreover, the arbitrator must render a decision within 48 hours of the hearing. A review of 31 published arbitration decisions in the steel industry indicates that where the expedited health and safety arbitration was employed, the average time from filing a grievance to the rendering of an arbitration decision was less than 10 days. A provision added in the 1977 Basic Steel negotiations authorizes local parties to make special arrangements for immediate arbitration of health and safety disputes (section 14.71 of the Basic Steel Agreement). For comparison, a 1974 Senate oversight committee reported that OSHA's average response time for employee-initiated complaints varied from 12 to 101 days among OSHA regional offices.[20]

According to information provided by the United Steelworkers, 186 health-and-safety-related grievances have been arbitrated in Basic Steel since 1966.[21] The union has prevailed in 73 of these disputes. Table 5.1 breaks down these decisions by subject and illustrates the relationship between health and safety and other terms and conditions of employment.

One consequence of providing a right to refuse hazardous work is that it encourages a union to characterize disputes that only collaterally raise health and safety issues as health and safety grievances. By doing so, the union gains the right to strike during the term of the contract over otherwise "nonstrikable" issues. For example, few issues threaten the security of a union more than proposed reductions in crew size. Such reductions typically mean fewer jobs for union members and more work for employees still on the job. In some cases reduction in crew size also creates additional hazards for workers. The difference in union success rates evident in Table 5.1 between crew-size arbitration awards (18 percent) and other hazardous-work decisions (47 percent) suggests that some Steelworker locals may be using the health-and-safety-dispute resolution mechanism to achieve non-health-and-safety-oriented goals. It should be noted, however, that misuse of the process in this manner is not necessarily bad. It creates a symmetrical incentive for management to pursue rapid resolution of all health and safety disputes. Thus, both sides are interested in quick arbitration of health and safety grievances—labor in order to ensure that workers will not lose wages while protesting hazardous conditions and management in order to minimize the cost of otherwise illegal strikes.

Table 5.1 also illustrates the narrow scope of inquiries by steel industry arbitrators into health and safety matters. An arbitrator sits only by express agreement of the parties, so he can only rule on matters the parties agree he should decide. Typically, this is limited to interpreting the provisions of the collective bargaining agreement. Since the duty of management to provide a safe workplace as defined by the collective bargaining agreement is narrow,[22] the arbitrator in steel considers only a limited range of issues in deciding health and safety questions: Is the hazard changed from the normal hazard inherent in the job? Did the worker comply with safety rules? Does the agreement require the company to provide a specific piece of safety equipment, such as protective shoes, gloves, aprons, etc.? The arbitrator does not consider general matters such as the maximum safe level of coke-oven emissions to which workers can be safely exposed, or whether the company should be required to institute stage charging as a protective measure. Thus, the arbitrator's scope of inquiry is constrained by the breadth of the substantive standard embodied in the collective bargaining agreement. It is also limited by his expertise in health and safety matters. Currently the steel industry employs regular arbitrators, with no special training, to hear health and safety disputes.

Table 5.1
Resolution of Health and Safety Grievances in Basic Steel

Subject of Grievance	Number of Cases	Union Success Rate
Refusal to perform work due to hazardous conditions	72	47%
Refusal to perform work due to reduction in crew size	39	18%
Dispute over provision of protective equipment or services	42	43%
Discipline for failure to comply with safety rules	15	27%
Other	18	55%
Total	186	73%

From Theory to Practice

The United Association of Plumbers and Pipefitters

The United Association is a craft union whose members work mainly in construction. It differs from both the United Auto Workers and the United Steelworkers because of special characteristics of construction labor. First, the tenure of a construction worker on any particular job is short. Unlike a factory worker, who is likely to be employed year after year at the same site, a construction worker moves on after completing his work on a particular construction job. Second, because of this short tenure, a construction worker will often work for a number of different employers at a number of different sites each year. As a result, both employer-employee and union-management relationships tend to be less rigidly structured than in other industries. Construction workers do not work under the same foreman for long periods of time, and there is no large job-site union bureaucracy such as is typically found in a steel or auto plant. Third, because of the nature of their work, construction workers generally exercise more discretion in their jobs than workers in most manufacturing industries. For example, plumbers must design and install plumbing systems tailored to meet the requirements of each individual job site. Construction workers tend to be highly skilled, and many learn their trades through sophisticated apprentice-training programs run by their unions. And fourth, construction work differs from other types of employment because construction workers rarely have a fixed work station at a particular job site. Unlike the factory worker, who usually stays at a single location within a plant, the construction worker moves around the structure under construction. These are not the only distinctions between construction and other industries—books have been written on the uniqueness of construction[23]—but these distinctions have significant implications for collective bargaining over health and safety issues in the industry.

The job mobility of construction workers affects their exposure to hazards. Over time, the factory worker will learn what hazards are peculiar to his work station, and how to avoid them. In contrast, the construction worker is exposed to a new and different array of conditions at each job. Unlike the factory worker, he generally does not have the advantage of being informed of the hazards by experienced fellow workers. Moreover, as he follows his work around the job site the construction worker is constantly exposed to hazards created by other workers in other trades, working for other subcontractors and represented by other unions. The plumber not only has to avoid creating hazards by his own carelessness; he must also consciously avoid the open wire being installed by the electrician next to him. Finally, because the workplace is itself being manufactured, the working environment at a construction site changes greatly over time.

Chapter 4 argued that collective bargaining offers a number of structural advantages over government-managed "command and control" regulation in limiting workplace hazards. This argument, however, is predicated on a continual union presence in the workplace that facilitates the recognition of hazards and the design and implementation of hazard-limiting procedures. In construction the union's presence on the job site is limited. Because the industry is organized by craft, there may be only one or a few members of a given union on a small job. The union presence on large jobs is also inhibited by the short duration of employment at the job site. One does not find a union hierarchy of zone and grievance committeemen in construction, because workers are not employed at a single job long enough to create such a bureaucracy. A union's presence at a construction site is often limited to occasional visits by the union business agent, an elected official who handles grievances and jurisdictional disputes, directs strikes and boycotts, and generally represents the interests of the rank and file in dealing with contractors.

For these reasons, construction unions seldom negotiate detailed local agreements for the abatement of specific hazards (as in steel) or agreements providing for union inspection of the workplace (as in the automobile industry).[24] Instead, health and safety provisions in construction focus on such issues as prohibition of unsafe tools, regulation of minimum crew size required to move heavy objects, hazard pay for working at heights, and amenities at job sites (potable water, toilets, warming sheds, etc.).[25] These agreements differ from agreements covering factory workers in that their primary purpose is not to improve the terms and conditions of employment of a specific group of workers currently employed by a particular employer, but rather to standardize the general terms under which labor is bought and sold.

The factors that inhibit negotiation of detailed hazard-abatement agreements and union inspection programs in the construction industry also inhibit enforcement of "command and control" regulation. Like the workers, OSHA inspectors are often unfamiliar with the unique characteristics of each construction site. Since they are only present for a short time, inspectors are unlikely to observe many hazards created during the changing course of a construction project. Moreover, the present system is ineffective in abating hazards created by workers themselves in the exercise of their professional discretion.

Efforts to control job hazards in construction, whether initiated by labor, management, or government, must recognize the atypical structure of employment in the industry. Because this structure does not lend itself to inspection by a single union representative or to regulation by government inspection, improving safety conditions requires altering the work practices and standard operating procedures of the workers to enable them to recognize and abate hazards on their own.

Fortunately, the necessary training can be included in existing apprenticeship programs. The United Association of Plumbers and Pipefitters has pursued this strategy in developing three health and safety training programs: a course (with a programmed-learning text) for all apprentices, a training program for teachers in the apprenticeship programs throughout the country, and a training program for union officers and business representatives.

The health and safety course is mandatory for all apprentices. It includes 12 hours of training in safe work practice, OSHA regulations and procedures, and recognition of the most common hazards in the plumbing and pipefitting industry. In addition, the United Association has integrated health and safety discussions into the training sessions for specific procedures such as welding, acetylene cutting, and rigging. Apprentices must pass a final exam on health and safety.

To be certified as an apprentice instructor, an individual must attend the annual Instructor Training Program at Purdue University for 5 years. The annual program consists of one week of intensive classroom instruction in teaching techniques as well as other topics covered in the apprenticeship curriculum. In 1976 the union added two courses in occupational health and safety to the instructor training curriculum: one covering general principles of job safety, OSHA standards and procedures that relate to the industry, common hazards encountered in plumbing and pipefitting, and general pedagogical instruction in the use of the programmed-learning text, and one course devoted entirely to principles of welding safety. Apprentice instructors who participate in the health and safety portion of the Purdue program for 2 years (40 hours) are authorized to issue Department of Labor–OSHA certificates to apprentices who complete the health and safety section of the union's apprenticeship program. The cost of the Instructor Training Program is borne by a trust fund established by the union and the National Contractors' Association; the expenses of individual apprentice instructors are paid by their local joint labor-management apprenticeship committees.

The one-week training program for union workers and business representatives is also offered at Purdue University. However, unlike the instructor training program, it is organized and run entirely by the union. Between 500 and 600 union officials attend the program each year, representing approximately half of the locals. For the past 2 years, 150 of the attendees have elected to take a 3-hour seminar entitled "Safety and Health Responsibilities of Local Union Officials," which covers the rights and duties of workers under the OSH Act, the procedures involved in filing an OSHA complaint, the subject of local bargaining over health and safety issues, and the activities of the union's Department of Safety and Health.

The objective of all three training programs is the same: to provide union members with sufficient knowledge so that they can both avoid creating hazards in their own work and exert pressure on employers to abate hazards within their control. To the extent that accidents are caused by worker carelessness, training programs like the United Association's are the only intervention instrument that can possibly lower the accident rate, short of an elaborate system of penalizing workers who exhibit careless behavior.[26] If workers learn to recognize violations of specific OSHA standards, they can exert pressure on management by making a credible threat to file a complaint that is likely to result in an OSHA inspection and sanction. From a policy perspective, training programs have the additional virtues of being cheap, posing little threat to managerial autonomy, and being available to unions without protracted contract bargaining.

Union Health and Safety Activities in Other Industries

Negotiated Changes
Union health and safety activity is not confined to the industries described in the preceding case studies. Of all collective bargaining agreements reviewed in a recent Bureau of Labor Statistics study, 93 percent contained some reference to health and safety.[27] However, relatively few of these agreements are as comprehensive as the above three.

The traditional labor-relations response to occupational hazards is formation of a health and safety committee. As table 5.2 indicates, these committees are most common in manufacturing. Health and safety committees vary substantially in the influence and authority they exercise within the workplace. Of the agreements referring to such committees, only 10 percent permit the committee to establish and implement safety rules and work practices independently. Most committees merely investigate accidents, review complaints of hazardous conditions, and make recommendations to higher management.[28] Occasionally, however, these advisory committees provide a forum for serious discussion and resolution of in-plant health and safety problems. The effectiveness of a committee is determined in large part by the degree of pressure exerted on management by OSHA and by the commitment of both the union and management to improving conditions.[29]

Only 73 percent of the 1,724 agreements reviewed provided for disclosure of job hazards to workers. These agreements cover 1.1 million workers, of whom 819,000 are employed in transportation equipment and primary metals manufacturing—industries dominated by the United Auto Workers and the United Steelworkers.[30]

Negotiated performance standards or engineering controls for the abatement of specific hazards are also rare, as table 5.3 indi-

Table 5.2
Prevalence of References to Health and Safety Committees Among 1,724 Collective Bargaining Agreements Reviewed by Bureau of Labor Statistics

	Industry	
	Manufacturing	Other
No. of agreements	908	816
No. of agreements referring to health and safety committees	407	169

Source: BLS bulletin (note 27), p. 51.

Table 5.3
Prevalence of Provisions for Protection Against Noxious Gases or Dust Among 1,724 Agreements Reviewed by Bureau of Labor Statistics

Provision	No. of Agreements	Workers ($\times 1,000$)
Ventilation	177	1,017
Respirators, dust masks	135	735
Time limit on exposure	10	46
Monitoring devices	44	1,041
Regulation on use, handling, storage of gas	16	70
Protection of unspecified kind	2	3
Total	285	2,218

Source: BLS bulletin (note 27), p. 58.

cates. Ventilation provisions are typically of the following kind: "Proper ventilating systems shall be installed where needed and maintained in good working order." [31] Similarly, respirator provisions require the company to furnish respirators to workers in specific jobs. The 10 agreements that specify time limits on worker exposure to noxious gases come closest to incorporating a negotiated standard into the collective-bargaining agreement; however, these cover only 46,000 workers. Monitoring devices are more common, appearing in about 15 percent of the agreements covering over a million workers. In effect, the presence of a monitoring device is almost as useful to a union as a negotiated performance standard. By recording the concentration level of a hazardous substance, the device makes union complaints to management credible and provides a basis for union-initiated OSHA complaints. The impact of the monitoring device is enhanced even further if there is an OSHA standard or threshold limit value for the monitored substance.

Although relatively few collective bargaining agreements contain OSHA-type health and safety standards, negotiated agreements concerning other aspects of the employer-employee health and safety relationship are common (see tables 5.4 and 5.5). For example, nearly half of the agreements stipulate that employees must comply with company safety rules. Pledges by management to comply with federal and state health and safety laws are contained in a slightly smaller proportion of the agreements. Safety-equipment provisions, which are common, establish whether labor or management is responsible for providing specific articles such as hard hats, steel-toe shoes, and gloves. Provisions mandating minimum crew size for specific tasks are most common in the construction and transportation industries. Though crew standards are couched in health and safety rhetoric, the motivation behind them is often preservation of union jobs. Sanitation clauses, which obligate management to provide sanitary facilities and drinking water, also are most common in construction, where such facilities often would not ordinarily be found at the job site. Many employers provide physical examinations as part of a health-care package. Often, they are preconditions of employment. Such exams are rare in construction, where employment is of short duration. Provisions relating to accidents describe the investigative procedures to be followed. First-aid clauses specify the minimum medical services that must be provided by the employer at the job site. In larger manufacturing facilities these often include a plant nurse or doctor, an ambulance, and a first-aid kit. Agreements governing a disabled worker's right to continued employment are common in manufacturing. Except in the utilities industry, they are rare in nonmanufacturing, where employment opportunities with a particular employer are typically less diversified than in manufacturing. Agreements providing for hazard pay are rare in every industry except construction and transportation, where workers must occasionally perform abnormally dangerous tasks (welding tanks, transporting dangerous cargo, etc.).[32] Half the construction agreements and a

Table 5.4
Incidence of Specific Health and Safety Provisions Among 1,724 Agreements Reviewed by Bureau of Labor Statistics

Provision	Industry Manufacturing	Other	Combined
Employer compliance with safety rules	30%	38%	34%
Employee compliance with safety rules	50%	43%	47%
Safety equipment	54%	44%	49%
Crew size	6%	20%	13%
Sanitation, housekeeping, hygiene	36%	42%	40%
Physical exam	29%	25%	32%
Accident investigation	28%	25%	27%
First aid	39%	17%	28%
Transfer rights for disabled workers	38%	11%	25%
Hazard pay	6%	25%	15%

Source: BLS bulletin (note 27), pp. 57–63.

Table 5.5
Incidence of Specific Health and Safety Provisions Among 181 Collective Bargaining Agreements in the Plumbing and Pipefitting Industry

Protective gear	87%
Compliance with federal or state safety law	45%
Shanties, weather protection, etc.	22%
Drinking water	22%
At least two men on hazardous work	22%
Sanitary facilities	20%
Safety committee	8%
First-aid equipment at job	8%
Safe ladders and scaffolding	8%
Shoring and cribbing of ditches	5%
Dead-man's switches	4%
Legal testing of gas lines	4%

quarter of the transportation agreements contain hazard-pay provisions.

Finally, in recent years a few large unions have negotiated research agreements with large employers. Typically these agreements allocate funds for long-term studies of specific industry hazards to be performed by university-based research teams. For example, the United Rubberworkers' agreement allocates a half cent per worker per hour for health-related research to be conducted at the Harvard University and University of North Carolina Schools of Public Health. The 1974 Basic Steel Agreement allocated two million dollars for research related to the health effects of employment in coke-oven plants. Similarly, in their 1976 negotiations GM and the UAW agreed to investigate the feasibility of jointly sponsoring a university study into the use and effects of lead in GM operations.

Non-Negotiated Changes

In addition to negotiating health and safety agreements with employers, many international unions also provide consultation services to their locals. Most often these include education programs, health surveys, hazard inspections (where permitted by the local bargaining agreement), and assistance in negotiating local hazard-abatement agreements. Though these services can be described, it is difficult to obtain meaningful measures of their prevalence. In 1976, however, the Ralph Nader–affiliated Health Research Group surveyed 15 major unions to determine the number of staff members each union employed in an occupational health capacity.[33] In all, the 15 unions employed 1 full-time and 2 part-time health-related professionals (hygienists, chemists, safety engineers, etc.), 7 full-time and 9 part-time other professionals (industrial engineers, lawyers, economists, etc.), and 14 full-time and 8 part-time union staff personnel. The United Steelworkers had the largest full-time health and safety staff among the unions surveyed, with 10 people. In contrast, the Steelworkers employ 14 people in their pension and insurance department, 24 in their public relations department, and 27 people in their legal department. As a further basis for comparison, all but 28 of the 2,000 physician members of the Industrial Medical Association are employed by management. The remaining

28 work in union clinics, performing general health care (not specifically job-related).[34] The data suggest that job safety and health does not command a relatively large commitment of resources on the part of the international unions surveyed.

Conclusions

An interesting pattern emerges from the three case studies and the statistics on health and safety bargaining: Little bargaining activity is directed at specific occupational hazards. Except in the case of the engineering controls required by the Clairton agreement, unions do not appear to engage in comprehensive efforts to negotiate OSHA-type standards. Instead, they concentrate on negotiating agreements that permit enforcement of standards set by others, establishing procedures for resolving health-and-safety-related labor-management disputes, obtaining agreements governing work rules and wage rates that are in some way sensitive to the existence of job hazards, and establishing training programs to make their members more safety-conscious.

Looking only at the content of collective bargaining agreements understates the true amount of union health and safety activity to some extent. Some informal bargaining over the existence of job hazards must occur during grievance procedures and during the deliberations of safety committees, but this bargaining often is confined to rather narrow safety issues, such as whether the exhaust fan in a paint-spraying booth is working properly. Whether unions engage in this informal hazard-limiting activity is determined by two factors: the ability of union personnel to recognize the existence of hazards and the priority given to health and safety by the union.

Although the data are not conclusive, the case studies suggest that the health and safety functions that are performed by the unions are performed well. As a means of enforcing existing OSHA standards, the union-representative approach of the GM-UAW agreement appears to be superior to random, infrequent OSHA inspection. The representative is in the plant every day, he knows the problems of the plant and is in a position to help implement compliance programs, and he can complain to OSHA if management is unresponsive to his demands. Experience under the GM-UAW agreement suggests that management compliance with OSHA regulations can be improved by encouraging unions to enforce OSHA regulations.

The dispute-resolution mechanisms that have evolved in the auto and steel industries are well suited to dealing with the health and safety disputes that arise in those industries. In particular, the expedited arbitration procedure in steel is a relatively swift method for determining whether a worker is justified in refusing to perform work that he believes to be excessively hazardous. Since this is a problem that arises frequently in many work settings and that OSHA is poorly equipped to deal with, OSHA should encourage labor and management to negotiate expedited arbitration procedures for resolving disputes of this type. However, arbitration is not a substitute for government regulation as a means for determining what constitutes an inherently hazardous environmental condition.

In some job settings (most typically in construction) where enforcement of regulations is complicated by the transience and changing nature of job sites, the work habits and job practices of the individual workers must incorporate hazard recognition and abatement. Since workers in the construction unions acquire their skills through union-directed apprentice programs, it is logical to channel regulatory resources through union health and safety training programs like those of the plumbers and pipefitters.

The scarcity of agreements directed at the elimination of specific job hazards is puzzling. Where hazards are not governed by externally defined standards, one might expect unions to take an active role in seeking to protect their members. While the Steelworkers local at Clairton did engage in such activity, other coke-plant locals did not. Similarly, there is great variation in degree, scope, and substance of the health and safety activities and negotiated agreements among unions and industries. To the extent that the federal government is interested in supplementing the existing regulatory structure by encouraging additional bargaining activity of the type described above, it is necessary to understand why labor and management do not undertake more of these activities voluntarily.

6
Influences on the Likelihood of Health and Safety Bargaining

Stimulating labor and management to address health and safety issues on their own is a subtle task. It requires an understanding of what motivates the parties to pursue health and safety issues as well as knowledge of the factors that inhibit additional bargaining in this area. This chapter uses the cases from chapter 5 and theories of labor relations to describe both the incentives and the disincentives labor and management face in bargaining over job hazards.

The rules and procedures formally articulated in a collective bargaining agreement are not the only rules governing conduct in the workplace. Management defines some policies without consulting labor, labor defines some policies without consulting management, government-imposed laws and regulation proscribe some actions and mandate others, and still other aspects of workplace behavior are controlled by habit and custom. Conceptually, the complex of rules governing the work environment represents the output of an industrial relations system.[1] The actors in the system consist of workers and their organizations, managers and their organizations, and government agencies. The form, the scope, and the substance of the rules that emerge from the interaction of these parties are influenced by many factors related to the environment of the industrial relations system. By examining these factors in detail, we can gain insight into why some unions commit major portions of their bargaining resources to the abatement of job hazards and why others do not.

Technological Factors

Since technology determines the range of occupational hazards found in any working environment, it also affects the motivation of the parties to establish rules governing hazards. Coke-oven workers are exposed to hazards from noxious gases because it is the essence of producing coke to produce dangerous fumes, and current technology can only keep a limited fraction of these from workers. As a result, rules relating to the control of coke-oven emissions are found in the steel industry. In contrast, there are few health and safety rules of any nature in the retail trade industry, because the technology generates few hazards. In addition to influencing the incidence of health and safety rules, technology also affects the form of the established rules. For example, there are thousands of different hazards distributed along the assembly line in an automobile plant, with workers at different stations exposed to different hazards. Since the potentially dangerous conditions are too numerous and too varied to be controlled by regulations or negotiated engineering controls, General Motors and the United Auto Workers negotiated an agreement that established a labor-management inspection team and gave it the power to identify and abate hazards. In contrast, the technology of a coke plant dictates that some type of engineering or performance standard either be negotiated by the parties or imposed by regulation. There is basically one large hazard in a coke plant; it affects virtually every worker and it can be controlled only by some form of explicit or implicit standard.[2] An inspection program alone cannot control coke-oven emissions. Technology also influences the form of health and safety rules because it determines the cost of alternative hazard control strategies.

Market Conditions

Market or budgetary constraints also affect the motivation of the parties to establish health and safety rules and the form of the rules themselves. Recent events in the automobile industry clearly demonstrate the impact of financial status and competitive position on the willingness and ability of management to implement expensive environmental controls. Similarly, labor is often unwilling to push hard for improvements in health and safety conditions in financially unstable industries. Workers in the secondary lead-smelting industry have not enthusiastically embraced proposed federal regulations that may force the closing of older, inefficient plants and cause loss of jobs. To quote an old saying from the steel industry, "Where there's smoke, there's jobs." The influence of market structure on the form of health and safety rules is more subtle. Disruptions in the supply of steel are extremely costly because steel is a primary material in other manufacturing industries. As a result, health and safety disputes are arbitrated in steel, and this eliminates the possibility that they could lead to expensive shutdowns.[3] In contrast, health and safety disputes in construction generally are not arbitrated, notwithstanding their costly impact (typically, they shut down the entire construction site); the market structure for construction services precludes establishment of the institutional relationships necessary to support arbitration.

Power Relationships

Some health and safety bargaining is motivated by the necessity to define the rights of the parties (the right to refuse hazardous work, the right to transfer following injury, the right to discipline for failure to comply with safety rules, etc.). At the plant level, the relative power of the actors determines, in large part, how health and safety issues are resolved. Some health and safety bargaining is purely distributive—what labor gains, management loses.[4] Consequently, strong unions can force management to bargain over health and safety issues and can win concessions at the bargaining table that weaker unions cannot. At the national level, the distribution of power between labor and management affects the composition of laws and regulations governing workplace health and safety. Politically powerful unions can occasionally obtain through regulation what they are incapable of obtaining through negotiation. The OSHA coke-oven standards forced the steel manufacturers to adopt operational changes that had been implemented at Clairton only after expensive negotiations.

Since federal regulation allows unions to husband scarce bargaining capital to achieve non–health-and-safety bargaining goals, it discourages some negotiation over health and safety issues.[5] Contrary to popular belief, however, OSHA regulation is not a free resource for labor. Labor devotes substantial lobbying efforts to maintaining the benefits of OSHA. More important is the fact that, to the extent that federal regulation forces management to incur large hazard-abatement costs, it diminishes the ability of labor to extract non–health-and-safety demands from management.

Preferences and Attitudes

Technology, market or budgetary constraints, and the distribution of power only define the environment in which health and safety

Influences on Likelihood of Bargaining

bargaining occurs. The motivation or willingness to bargain to achieve improved health and safety conditions is determined ultimately by the preferences of the parties. In every bargaining situation, a union has at its disposal only a finite amount of bargaining capital. This capital derives from the union's ability and willingness to interrupt management's supply of labor. Like any other kind of capital, however, bargaining capital has its opportunity cost. It may be expended to achieve higher wages, more job security, better pensions, or improved health and safety conditions. A union that chooses to pursue one bargaining goal necessarily does so at the expense of others. Consequently, the priority given to health and safety objectives relative to other bargaining goals determines, to a large extent, whether health and safety provisions find their way into collective bargaining agreements.

To say that workers are not interested in occupational safety and health is to oversimplify greatly. Obviously workers are interested in safer working conditions, but at what cost? Coke-oven workers prefer employment in a health-threatening environment to no employment at all, because they could choose to be unemployed. What is not clear is the extent to which they would willingly forgo wage gains and other benefits to clean up their workplaces.

A number of researchers have attempted to assess the intensity of worker preferences for improved working conditions through attitudinal surveys. These studies are somewhat suspect, because respondents often give answers that do not necessarily reveal their true preferences but rather indicate what they wish to be heard saying. Since the strategic answer is typically the answer that places the highest possible valuation on safety, studies of this type may overstate the degree to which workers value improved health and safety conditions. Nonetheless, they are useful in that they provide some insight into how workers might respond when faced with perfect information and a simple binary choice. Steven Kelman surveyed 383 members of the Oil, Chemical and Atomic Workers Union. When he asked whether they would prefer to receive a $5,000 after-tax payment in lieu of elimination of exposure to a known occupational carcinogen, 89 percent of the workers preferred elimination of the hazard to the money. In another question,

64 percent of the workers said they would refuse to take a job that offered a one-in-a-hundred chance of serious injury each year in return for a 15 percent after-tax wage increase. Other attitudinal surveys also suggest that workers value safe working conditions highly relative to other bargaining objectives.[6] Observation of worker behavior in the marketplace, however, provides some indication of the size of the strategic bias. The United Steelworkers were forced to renegotiate the incentive pay schedule to make the Clairton agreement acceptable to the rank and file, because the workers were unwilling to forgo incentive income in return for a cleaner work environment. The amount at issue was approximately $1,000–$2,000 per worker per year in pretax earnings.[7] As both theory and the Clairton experience suggest, ultimately the intensity of worker preferences acts as a constraint on the negotiation of hazard-abatement agreements. The magnitude of the response to the Kelman survey, however, does not lend credence to the theory that there are few negotiated hazard-abatement agreements simply because workers are not interested in safer working conditions. Even if the survey overestimates a worker's willingness to pay for hazard abatement by a factor of 2 (that is, if 89 percent of workers would forgo $2,500 rather than $5,000 in after-tax earnings), a 100-member local should still be willing to forgo $250,000 in wage gains to support elimination of a known carcinogen.[8] If the surveys are to be believed, workers value their health sufficiently to justify substantial investments in improved health and safety conditions. Worker ambivalence alone cannot explain the low incidence of negotiated hazard-abatement agreements.

Management attitudes toward the control of occupational hazards also influences the level of substantive bargaining. However, it is again an oversimplification to allege there are not more negotiated health and safety agreements because employers are not interested in the welfare of their employees. As noted earlier, even cynical employers care about the welfare of their workers, at least because accidents and illnesses impose costs in the form of production losses, compensating wage premiums, and higher insurance payments. Moreover, most employers have a sincere desire to see that their workers avoid injury. In fact, a recent survey of managers concluded that there was less conflict over health and safety than over any other collective bargaining issue.[9] In practice, three factors determine management's willingness to negotiate any kind of health and safety agreement: the net cost of the agreement (costs minus savings from avoided production losses, reduced hazard pay, etc.), the expected cost of nonagreement, and the extent to which the hazard abatement infringes upon managerial autonomy. Expensive hazard-abatement programs almost always impose net costs on management. If they did not, management would undertake them without prodding from labor. Where proposed health and safety measures impose net costs, management will only accede to union demands if the expected cost of nonagreement exceeds the net cost of the union proposals. Thus, management's willingness to bargain over health and safety issues can be increased by either reducing the net cost of the union proposal or increasing the expected cost of nonagreement, which in turn is determined by the union's ability and willingness to be intransigent or to engage in strikes or slowdowns to achieve health and safety goals.

Finally, some health and safety initiatives are opposed not because of their financial cost but because they constrain managerial discretion. Perhaps no other tenet is held so strongly or so uniformly by managers as the principle that "management manages and labor works." Bargaining over health and safety issues threatens managerial autonomy particularly because it involves labor in capital allocation, an area previously within the exclusive domain of management. Moreover, bargaining over the dimensions of a safe work practice (such as the number of people required to perform a job safely) interjects the union into day-to-day managerial decisions. General Motors offered this rationale for its initial opposition to the UAW health-and-safety-representative proposal.

In summary: The preferences of workers should determine the motivation of unions to seek improvements in health and safety conditions. The cost of labor's proposals, labor's ability and willingness to engage in coercive activities in support of the proposals, and the degree to which the proposed changes infringe upon managerial autonomy shape management's response.

Availability of Information

Lack of information can act as a barrier to bargaining over health and safety in at least three different ways. First, workers may simply not know that a particular process, task, or chemical is hazardous. If workers are unaware of a hazard they will not bargain to get rid of it, notwithstanding strong preferences for safe working conditions. Moreover, it is difficult to make a credible case at the bargaining table (or in the union hall) for the abatement of a suspected hazard without objective evidence that the condition is actually dangerous. For example, coke-oven emissions did not become a serious collective bargaining issue until the completion in 1966 of a study of lung-cancer mortality among Allegheny County coke-oven workers. Not until the study was updated and published in the *Journal of Occupational Medicine* in 1972 was the magnitude of the problem fully appreciated.

Second, bargaining may be inhibited by a lack of knowledge of the level of a particular hazard. A union may know that a substance is potentially hazardous but be incapable of determining whether it is present in sufficient quantities to pose a risk to human health. Part of the reason the United Steelworkers did not negotiate a performance standard for coke-oven emissions was because they lacked the expertise to monitor compliance with the standard. The United Auto Workers' approach to dealing with hazards is effective because it gives the union the technical capacity to independently observe intermittent hazards such as noise.

Third, bargaining over possible engineering controls is occasionally inhibited by a lack of knowledge of cost-effective control mechanisms. The parties may be able to monitor a known hazard, but may not know what to do to get rid of it. Typically, only labor will be lacking in this knowledge. Labor may even have trouble hiring experts—during the Clairton negotiations all of the recognized experts in the field of coke-oven design worked for the industry.[10]

Representation of Safety-Conscious Workers

Although strong rank-and-file preferences for safer working conditions and information about hazards and their control mechanisms are necessary for union health and safety bargaining, they are not sufficient conditions. For unions to take an active role in hazard abatement, their bargaining agendas must accurately reflect the preferences of safety-conscious workers. Unions formulate their bargaining goals and priorities through a political process that is not well understood. For any given issue, there exists a distribution of preferences among individual workers. For example, during the Clairton negotiations some workers valued possible gains from engineering controls very highly while others did not. Through consultation, persuasion, and compromise, union leaders gather these individual preferences into a bargaining agenda. Like most political processes, the formulation of a bargaining agenda is not equally responsive to all constituents. Some interest groups carry more weight than others.

There are a number of reasons to believe that workers with strong preferences for improved health and safety conditions might be systematically underrepresented at the bargaining table. First, only small groups of workers tend to be exposed to specific hazards that threaten personal safety through traumatic injury. For example, only crane operators and individuals who work in the vicinity of cranes are threatened by faulty hoisting equipment. In contrast, health hazards often threaten all the workers within a plant. Coke-oven emissions, vinyl chloride, asbestos, and cotton dust are good examples of health hazards that permeate the workplace, affecting all within to some degree. There is relatively little political return for union leaders who negotiate agreements for the abatement of specific safety hazards, especially outside the craft unions. Moreover, since workers are highly conscious of their relative positions within the workplace, agreements that benefit different groups within a union differently are politically unpopular.[11] As a result, safety hazards are rarely bargained over directly during contract negotiations. When union leaders want to be responsive to individual safety problems, they establish procedures for resolving safety disputes that appear to benefit all members of the union equally, such as safety committees and special arbitration procedures.

Second, union leaders generally are selected from the ranks of the higher-paid, more skilled members of the bargaining unit.[12] They tend to be older workers with considerable experience in union politics. In contrast, the workers most likely to be injured in an occupational accident are younger and less experienced.[13] Younger workers and minorities hold a disproportionately high percentage of the most health-threatening jobs within many workplaces. To the extent that workers in high-risk positions are underrepresented in the intraunion political process, the union's bargaining agenda will not reflect the preferences of the workers most likely to gain from improved health and safety conditions. (The assumption that workers at risk are more health-and-safety-conscious than their fellow workers, while plausible, is difficult to test empirically.[14])

Third, unlike wage gains, the benefits from improved health and safety conditions are not always immediately apparent to the rank and file. For example, textile workers often suffer from a lung condition known as byssinosis, caused by the inhalation of microscopic particles of cotton dust. Because the particles are so small, workers often cannot directly observe increases or decreases in exposure levels. Because of the characteristics of the disease, the impact on worker health of a reduction in exposure level is not obvious for years. Moreover, it often takes years simply to implement a negotiated hazard-abatement program, in contrast with other types of collective bargaining agreements. As a result, the union leader who forgoes other benefits to negotiate reductions in worker exposure to hazards often has little to show for his efforts during the term of his office.

Cost to the Union

There are other costs associated with bargaining over health and safety issues. Disputes over compliance with contract agreements usually are settled through arbitration, the cost of which is shared equally by labor and management. Consequently, to enforce a negotiated health and safety standard a union must be prepared to incur substantial out-of-pocket costs. Since bargaining also occurs within the grievance process,[15] enforcement of health and safety standards through the grievance-and-arbitration procedure also forces a union to expend bargaining capital that might otherwise be used to favorably resolve other grievances. In contrast, if the same abatement procedures are incorporated in an OSHA regulation a union can obtain enforcement merely by filing an OSHA complaint. Consequently, from the perspective of labor, negotiated agreements are not only more expensive to obtain than OSHA regulations; they are also more costly to enforce.

Bargaining over job hazards also is expensive because it makes unions potentially liable for failure to enforce their negotiated agreements. Under existing law a union has no legal duty to negotiate any kind of health and safety agreement, so a union is not liable for an injury to a member where the union has assumed no duty to protect the member from injury. If, however, the union has negotiated an agreement that permits the union to inspect the workplace for hazards, an injured worker may claim that the union is liable in tort for his injury because it failed to discover and abate the hazard. From the perspective of the plaintiff, suing his union is desirable because it provides an additional source of recovery beyond the maximum available from his employer through workmen's compensation.[16] To date there have been only a handful of such cases, and most have been unsuccessful because of the heavy burden of proof imposed upon the worker-plaintiff.[17] Nonetheless, the threat of liability is sufficiently serious that a number of unions have privately negotiated agreements that provide for employer indemnification of the union in the event that the union is successfully sued by a member for failing to discharge its duties under a negotiated health and safety agreement.[18] More important, since unions can avoid liability entirely by not negotiating a comprehensive health and safety agreement the threat of liability currently discourages union activity in hazard regulation.

Role of Technical Consultants

Most collective bargaining agreements are negotiated by a small group of the principal officers of the union and the company. Labor is typically represented by the union president and a bargaining team elected from the other union officers. Management usually is represented either by the chief operating officer or (in larger companies) by the vice-president for labor relations. The company treasurer or comptroller is also usually a member of the bargaining team. What is significant about this array of parties is that nearly all of the bargaining over contract issues takes place between line officers representing both labor and management. In the case of established unions, the parties often know each other from previous negotiations. Staff assistants and technical experts play very minor roles in most contract negotiations; the actual decisions and negotiations are handled by the principals. Usually the lawyers and the technicians are called in only to answer questions and to put the agreement in writing.

Bargaining over health issues is complicated by the technical nature of the subject matter. Rarely do the principals have sufficient expertise to formulate and explore negotiating positions without substantial assistance from experts. For example, during the Clairton negotiations the United Steelworkers' bargaining team regularly sought the assistance of experts at the University of California to interpret management proposals and make union counteroffers.[19]

Though experts provide both sides with much needed help, they also inhibit the give-and-take that constitutes the heart of collective bargaining. With few exceptions, technical consultants do not appreciate either the political subtleties of negotiation or the relationship between health and safety rules and other terms and conditions of employment. Nor do they share the mutual understanding that often develops between principals who must live under the agreement that they negotiate and must look toward future negotiations. Consequently, bargaining over the abatement of specific job hazards is more difficult than other contract bargaining because negotiating teams often must cede substantial responsibility to technical experts unskilled in the art of collective bargaining and without ongoing responsibility.

The technical nature of health and safety issues complicates contract administration as well as negotiation. Even where parties possess sufficient expertise to negotiate an agreement, they often lack the technical sophistication necessary to implement it. Collective bargaining agreements are implemented in large part through the grievance-arbitration process. Company departures from contractually defined norms are remedied through grievances. Workers who fail to fulfill contractual obligations are disciplined, and the validity of disciplinary measures may be challenged by grievance. As the GM-UAW case illustrates, where grievance-arbitration procedures exist most grievances are settled short of arbitration by mutual agreement of the parties. For disputes arising out of interpretation of a negotiated health and safety agreement, however, the parties (foremen, personnel directors, grievance committeemen, and union officers) rarely have the technical sophistication necessary to make a reasoned assessment of a grievance. The United Steelworkers did not negotiate a performance standard at Clairton because they had no way of resolving the inevitable grievances that would have arisen from implementing such a standard. They were forced to negotiate engineering controls that could be easily observed. Similarly, the GM-UAW health and safety program has been successful because it provides the parties with the technical expertise necessary to resolve everyday workplace disputes. Most significant, the GM-UAW approach concentrates this expertise where it is needed most: at the plant level.

There is one other technical barrier that inhibits negotiation of sophisticated health and safety agreements: Like the parties, arbitrators also lack the technical expertise to resolve health and safety grievances judiciously. Most arbitrators are either lawyers, college professors, or labor-relations specialists, and have virtually no training in hazard recognition or abatement. A number of General Motors officials have said that one of the reasons the company refuses to negotiate companywide health and safety standards is that it does not trust an untrained umpire to enforce such standards correctly. And in steel, where health and safety disputes are arbitrated, the scope of inquiry of the arbitrator is very narrow precisely because arbitrators lack the background to make judgments more sophisticated than "whether the hazard is changed from the normal hazard inherent in the job." Because an arbitrator must be trusted and accepted by both labor and management, substitution of health and safety professionals (who might be viewed as having a protectionist bias) for existing arbitrators is not feasible.

Philosophical Factors

Bargaining over occupational hazards is constrained not only by the technical nature of the subject matter but also by the way labor and management think about health and safety issues. For many years labor has argued that workers have a right to work in an environment free from recognized hazards. "A worker should not have to choose between his life and his job," to quote an old labor expression. This philosophical position colors both labor's and management's views of health and safety bargaining in a number of ways.

The view that occupational health is a right necessarily implies that employers have a correlative duty to eliminate all conditions that may threaten the welfare of the labor force. Since logically one should not have to pay for what one is rightfully entitled, this philosophical position discourages labor from seeking abatement of job hazards through collective bargaining.

Labor's view that job safety and health is a right also logically implies that every worker has an identical claim to that right. In other words, labor would argue (and did in the congressional hearings preceding enactment of OSHA) that for workers in different work settings to be exposed to different levels of risk is inherently inequitable. Southern textile workers should be exposed to the same level of cotton dust as Northern textile workers; rubberworkers should be exposed to the same maximum noise level as steelworkers. In every case this level should be the "safe" level. Consequently, labor's preference for federal regulation of job hazards represents a means of standardizing worker exposure to job hazards nationwide, a result that would not be reached if hazards were abated through collective bargaining. The desire for standardization through federal regulation, however, goes beyond labor's philosophical position on worker entitlement to job safety. Labor has argued that employers whose plants are union-organized resist bargaining over health and safety issues because it places them at a competitive disadvantage relative to unorganized employers,[20] and that federal regulation is needed to overcome this competition-motivated disparity in willingness to bargain. This position is on its face disingenuous. All union gains obtained through collective bargaining place organized employers at a competitive disadvantage relative to unorganized employers, yet labor does not urge federal regulation of all terms and conditions of employment as a means of overcoming management's reluctance to bargain. In fact, labor's preference for federal regulation is more likely motivated by a desire to avoid a net transfer of business and jobs from organized to unorganized employers resulting from price increases mandated by negotiated health and safety measures.

Labor's position that health and safety problems are primarily the concern of management has encouraged labor to focus on collateral aspects of health and safety issues. Rather than negotiate direct hazard-abatement agreements like the Clairton agreement, unions have concentrated their efforts on establishing safety committees, obtaining improvements in plant first-aid facilities, negotiating for management provision of safety clothing, securing the seniority rights of workers transferred because of injury, and obtaining pay for time spent on accident inspections and safety committee work.

Finally, labor's philosophical position has also influenced management's response to union collective bargaining initiatives. Where convenient, management has adopted the view that health and safety is the exclusive responsibility of management in order to thwart labor's attempts to assume a more active role in day-to-day hazard identification and control. This was the basis for General Motors' initial opposition to the UAW health-and-safety-representative proposal, and it remains the basis for GM's opposition to arbitration of health and safety disputes.

Conclusions

The existence of OSHA discourages some bargaining over safety and health issues by offering a relatively cheap alternative to negotiated hazard abatement. Health and safety issues do not command a high position on union bargaining agendas because there is little political return on cleaning up the workplace; changes are often not recognized for years and the individuals most likely to benefit tend to be underrepresented. In addition, unions prefer standardized regulations to negotiated agreements that may place organized employers at a competitive disadvantage. When unions are inclined to pursue hazard abatement, they often lack the necessary technical expertise. To assist in enforcement of existing regulations, unions must be knowledgeable about the specifics of relevant regulations and be capable of determining when management is not in compliance. Abatement of hazards through union-initiated bargaining requires even greater technical sophistication. The union must be capable of recognizing a dangerous condition, it must know approximately the relationship between exposure and risk, and it must understand feasible control mechanisms. Successful bargaining requires that much of this knowledge be possessed by both the negotiators and the arbitrators who will enforce the agreement. Union health and safety initiatives must also overcome management's reluctance to cede authority to labor for even partial management of health and safety issues.

7
Ways to Encourage Labor-Management Health and Safety Activity

Collective bargaining is not a substitute for government regulation of job hazards. Scientific complexity renders negotiated standard setting for health hazards particularly impractical. Nonetheless, organized labor has the capacity to perform a number of quasiregulatory functions that, if conducted on a large scale, would significantly improve the performance of the federal regulatory effort.

Specifically, unions are capable of pressuring management to comply with existing standards. This pressure can take many forms, ranging from gentle persuasion to slowdowns and strikes. Though union pressure will not ensure perfect compliance, it can dramatically supplement the minimal compliance incentives provided by the present system of inspections and fines. Moreover, unlike the threat of inspection, union pressure is immediate, potentially constant, and focused. Labor can also bargain directly for the abatement of some safety hazards. A union's familiarity with the workplace should give it an advantage over OSHA in dealing with intermittent and workplace-specific hazards. In addition, labor-management apprenticeship programs provide an opportunity for OSHA-specified health and safety training to be integrated into a worker's training. Finally, the arbitration system is a potential forum for quick resolution of disputes over an employee's right to refuse hazardous work.

If OSHA is to capitalize on these opportunities it must encourage labor to take a more active role in making the workplace safe. To date, OSHA's direct efforts to stimulate more union involvement in health-and-safety-related activities have focused almost exclusively on training programs. The Office of Training and Education funds a number of projects at major universities aimed at educating workers in OSHA-mandated rights and procedures, the techniques of hazard identification and abatement, and possible collective bargaining approaches to resolution of health and safety disputes. Unfortunately, labor training programs alone will not stimulate much health and safety bargaining activity. If labor and management are to be more active in making the workplace safe, government must find ways to address all of the barriers to collective bargaining over health and safety issues described in chapter 6. In addition to making more information available to unions, government must make health and safety bargaining politically attractive to union officers, lower the costs of negotiation to both labor and management, and address the technical and philosophical barriers to negotiation. There may be different opinions about the optimal strategy for achieving these objectives, but any policy designed to stimulate health and safety bargaining must address each of these problems.

This chapter describes one possible government strategy for increasing the involvement of labor and management in regulation of job hazards. The recommendations are grouped according to the activities unions would perform: enforcing OSHA regulations, identifying and abating hazardous conditions not covered by existing regulations, resolving health and safety disputes, and training workers. Table 7.1 summarizes the recommendations and the problems they are designed to address.

Promoting Enforcement of OSHA Regulations by Unions

Training Union Stewards

OSHA's response to public criticism of the efficacy of its enforcement procedures has been to upgrade and professionalize the compliance force. According to the 1977 Status Report on OSHA submitted by former Assistant Secretary Morton Corn, the agency has placed "major emphasis . . . on hiring professional health and safety compliance officers and training those currently employed . . . to bring them up to an acceptable level of performance."[1] Accordingly, OSHA has initiated a 3-year training program for health compliance officers, a 4-week course for new safety officers on OSHA standards and the hazards they cover, and a 2-week health course for current safety officers who lack training in health-hazard recognition. The agency is also busy trying to increase the size of its compliance force. Though these efforts are noble, they probably will not result in significantly greater employer compliance. As noted in chapter 3, doubling the size of the inspection force only increases the expected cost of noncompliance from $3 to $6 per year. Some of OSHA's inspector-training funds might be better spent by deprofessionalizing the

Table 7.1
Summary of Recommendations and Their Objectives

Recommendation	Objective
Train health and safety stewards.	Provide technical expertise in recognizing hazards and monitoring compliance with relevant standards. Strengthen union credibility in H & S bargaining. Create union H & S lobby.
Exempt workplaces from OSHA inspection.	Give management incentive to bargain with union. Give union leverage in obtaining walkaround inspection rights. Encourage cost-effective enforcement.
Respond quickly to union complaints.	Force management to take union demands for compliance seriously.
Exempt unions from liability	Lower perceived cost of bargaining over H & S issues.
Fund industry morbidity and mortality studies.	Draw workers' attention to risks. Raise H & S on union bargaining agenda. Give credibility to labor H & S demands.
Expand health-hazard evaluation program.	Provide labor with technical expertise. Strengthen union bargaining positions on H & S issues.
Expand right to refuse hazardous work.	Remove workers from risk. Provide more intraunion political power to safety-conscious workers.
Give H & S training to arbitrators.	Provide technical expertise necessary to implement negotiated H & S agreements.
Underwrite cost of H & S arbitration.	Lower cost of bargaining over H & S issues relative to OSHA enforcement.
Integrate H & S training into apprenticeship programs.	Develop safe work practices.

compliance force and training additional union health and safety stewards.

From OSHA's perspective it makes sense to train union health and safety stewards for a number of reasons. They cost OSHA nothing more than the training expenses. They provide unions with the technical expertise needed to make informed judgments about the existence of hazards and the extent of management noncompliance, and this increases the union's credibility on health and safety issues, which in turn strengthens its bargaining position. As the GM-UAW case illustrates, union stewards can provide the continuous and immediate compliance pressure that is missing from the present inspection-and-citation system of enforcement. If the training is focused on the particular hazards of the steward's work environment, a steward can be taught to do much of the work of an OSHA inspector in a single 40-hour course. By comparison, new OSHA safety inspectors receive 160 hours of general classroom instruction covering virtually all OSHA standards for all industries.[2] What a union health and safety steward lacks in technical sophistication is compensated for by his presence in and his familiarity with the workplace. In addition to providing compliance pressure, training one person in each workplace to be a quasiprofessional health and safety inspector also provides a basis for a health and safety lobby within each union. Furthermore, the safety steward gives the union the capacity to negotiate agreements whose implementation requires a technical monitoring capability on the part of the union.

Exempting Workplaces from OSHA Inspection
Training alone does not ensure that union personnel will either enforce OSHA regulations or independently pursue the abatement of job hazards. To function effectively as a quasi-inspector, a union steward must have as much access to the plant as an inspector. He must be free to walk around the work site, investigate complaints, and discuss health and safety issues with foremen, supervisors, and workers. It is a major failing of the present OSHA training program that it does nothing to encourage employers to give union health and safety representatives responsibilities commensurate with their training. OSHA presently trains a handful of health and safety rep-

resentatives and turns them loose in the hope that their unions will have the foresight and bargaining strength to negotiate agreements securing walkaround rights and hazard-monitoring equipment. Where unions fail to negotiate these provisions, the health and safety representative is mostly wasted. Where unions have negotiated such agreements, OSHA has ignored this information in targeting its scarce enforcement resources.

To encourage management to give union health and safety representatives responsibilities commensurate with their training, OSHA should exempt employers that meet specified conditions from all non-complaint-initiated inspections. A very large employer might be required to fund a full-time health and safety representative with daily walkaround rights; in a very small workplace the safety steward might devote only 4 hours per week to health and safety matters and have weekly walkaround rights. The time required of the health and safety representative could also be scaled to either the nature or the extent of the hazards. Similarly, employers might be required to supply the union with hazard-monitoring equipment before being exempted from inspection. Exemption should be on an experimental basis, as no empirical data exist to document the impact of union health and safety stewards on the incidence of hazards in the workplace. To guard against worsening of conditions in exempted firms, OSHA could require each such firm to demonstrate that its accident and illness rate has improved before renewing the exemption.

Exempting firms that satisfy the stated conditions from regularly scheduled OSHA inspections would give unions leverage in obtaining the rights requisite for an active role in health and safety matters. It would strengthen the bargaining position of unions on health and safety issues in a way that would make it difficult to trade off the newly obtained bargaining chips to achieve other objectives, such as higher wages. Exemption would offer the employer the opportunity to deal with his union rather than with a government agency that is sometimes perceived as irrational and unyielding. Also, delegating enforcement responsibilities to unions through exemption would allow OSHA to concentrate its scarce inspection resources on nonunionized workplaces.

Responding Quickly to Union Complaints

By now the careful reader will be asking, "But what if the union health and safety steward is incapable of pressuring management to comply with OSHA regulations?" Clearly, the relative bargaining strength of unions and management varies from firm to firm. Some unions would find it easier to enforce regulations than others. Since exemption can only be initiated with the agreement of the union, weak unions would not elect to participate in the program. Where management was recalcitrant on particular issues, OSHA could strengthen the bargaining position of the union by responding quickly and decisively to the steward's complaints.[3] A number of GM safety directors have reported that UAW health and safety representatives used the threat of a complaint to improve their in-plant bargaining position. If these threats are credible (that is, if management perceives that OSHA is responsive to such complaints), then the imposition of the threat alone should induce some firms to comply without OSHA inspection. If an employer refused to comply, then the union would still have the option of calling in OSHA for an inspection, and thus would be no worse off than under the present system.

Why should an employer participate in a program that places additional pressure on him to comply with federal regulations? The answer lies in the discretion enjoyed by the union in enforcing regulations. Unlike OSHA, a union can wink at *de minimus* violations that have little if any impact on worker safety and health. Given this discretion, a union will enforce standards selectively, scaling compliance pressure according to the degree to which the union cares about the hazard at issue. Some inconsequential regulations may be ignored entirely. Furthermore, the union enjoys greater discretion than an OSHA inspector in determining what constitutes compliance. In effect, exemption would free management to look at alternatives to the hazard-abatement solutions mandated by OSHA regulations. If a union is enforcing the regulations, any hazard-limiting procedure that survives the scrutiny of the union constitutes compliance.[4] Also, management has greater leverage in securing the participation of labor in implementing health and safety procedures if the union is responsible for the enforcement mechanism.

Exempting Unions from Liability

Chapter 6 noted that fear of liability may inhibit some unions from being more active in health and safety. Conceptually, the legal position of a union is similar to that of a physician who happens across an unconscious accident victim. The physician may be humanely inclined to treat the victim but be reluctant to do so for fear of malpractice liability. If society is interested in having the physician treat the injured person, it has two choices: legally compel him to treat all injured people he finds lying in the streets and subject him to some sanction if he fails to do so, or remove the disincentive to treat by exempting the physician from liability for everything except gross negligence. Since it is virtually impossible to specify the circumstances under which a physician might have a legal duty to initiate treatment, Good Samaritan statutes typically provide some form of exemption from liability. The problem posed by fear of liability on the part of unions can be addressed similarly: by imposing liability for failure to negotiate and enforce health and safety agreements, or by statutorily removing the threat of liability.

It would be far easier to encourage the desired outcome by exempting unions from liability than by compelling unions to enter into negotiated hazard-abatement agreements. Unions cannot be so compelled, because it is not possible to order parties to negotiate an agreement without first specifying the terms of the agreement they are to "negotiate." In effect, the mandatory agreement would constitute nothing more than a standard. Moreover, it is impractical for political reasons to impose liability on unions for failing to enforce existing standards. Labor has successfully resisted efforts to subject individual workers to OSHA fines in cases where workers are responsible for violation of OSHA standards (such as failure to use required protective equipment). Clearly, if labor can defeat a proposal to fine workers when they are at fault, then labor can defeat a proposal to fine unions that fail to exert pressure on management when management is at fault. Consequently, exempting unions from liability is the only feasible means of neutralizing the disincentives created by the threat of suit by a worker.

In practice, such an exemption would take away few rights from the injured workers who might be inclined to sue their unions. As noted earlier, workers must at present demonstrate breach of the duty of fair representation to recover in an action against their union. This would remain unchanged under union exemption from liability. All the proposed exemption would do is remove ambiguity in the present law concerning the potential liability (based on a negligence theory) of unions for failure to enforce either negotiated or existing health and safety standards. Injured workers would still be covered by their employers through workmen's compensation. This exemption would make explicit something that has been implied for years: that the employer is legally responsible for injuries to his workers and that a union is liable only when it contributes to the injury by acting in an arbitrary, discriminatory, or openly hostile manner toward the injured worker.[5]

Taken together, the first four recommendations constitute a coherent program for encouraging unions to actively assist in the enforcement of OSHA regulations. The first recommendation gives unions the technical capability necessary to recognize conditions of noncompliance, the second encourages employers to give union health and safety stewards the rights they need to do their jobs effectively, the third strengthens the hand of unions seeking to encourage management compliance, and the fourth removes a small but significant obstacle to greater union involvement. All but the last recommendation can be implemented without further legislation.

Promoting Negotiation of Extended Health and Safety Agreements

Getting labor more involved in the identification and abatement of hazards not covered by OSHA regulations is more difficult than simply encouraging labor to enforce existing regulations. The obstacles to negotiation of individual hazard-limiting agreements are formidable. For many hazards there is great scientific uncertainty over what constitutes safe or hazardous exposure. Unions often lack the information and expertise necessary to bargain intelligently over control mechanisms. Unions are not organized to be politically responsive to health and safety problems. The role of experts and staff complicates health and safety bargaining. The prospect of future OSHA regulation discourages bargaining. Philo-

Encouraging Labor-Management Activity

sophical views inhibit negotiation on both sides of the bargaining table, and in some cases unions simply lack the clout or the disposition to win expensive health and safety measures from management. Typically, these obstacles inhibit bargaining over complex health issues more than they affect safety negotiation. Accordingly, collective bargaining holds more promise as a means of controlling safety hazards.

As will be apparent from the discussion that follows, government's ability to encourage labor and management to negotiate over job safety and health is limited. OSHA has few carrots and even fewer sticks for altering the bargaining behavior of the parties. The strategies outlined in this section are designed to induce marginal changes to make the bargaining environment slightly more conducive to negotiation over health and safety issues. None of the recommendations that follow will totally eliminate any of the obstacles described above.

Funding Industry-Specific Morbidity and Mortality Studies

Under the division of labor articulated in the Occupational Safety and Health Act of 1970, NIOSH is responsible for conducting research related to occupational hazards. NIOSH also develops the criteria documents that provide much of the scientific basis for OSHA regulations. Not surprisingly, NIOSH coordinates its research with its development of criteria documents; if a criteria document is to be prepared, a matching study of worker exposure to the hazard is usually conducted. Two consequences of this coordination are that if no standard is contemplated no study will be performed and that, because the vast majority of criteria documents relate to specific occupational hazards, very few industry-specific studies are performed.[6]

Morbidity and mortality studies stimulate health-and-safety-related bargaining by drawing workers' attention to job-related risks and motivating workers to identify unknown dangers.[7] They also strengthen labor's bargaining position by providing a factual basis for claims about the impact of exposure to job hazards. Consequently, such studies can stimulate collective bargaining even when a standard is not contemplated. Because NIOSH does not consider the positive impact of morbidity and mortality studies on collective bargaining, it underinvests in this type of research. If government is to fund additional studies as a means of encouraging bargaining, the studies should be directed at particular industries or crafts and not individual hazards. Industry- or craft-specific studies have more impact on union bargaining priorities than hazard-specific studies because they highlight the cumulative and interactive effects of job hazards on the workers.

Firm-specific information also stimulates unions to bargain over health and safety issues. Workers and their unions are acutely sensitive to differentials between firms concerning terms and conditions of employment. Such differentials commonly play a large role in contract negotiations as well as union organization campaigns. For example, unions are quick to point out to employers that their workers earn less than the industry average, receive lower-than-average pensions, or have below-average medical benefits. Workers who know that their terms and conditions of employment are below average generally ask their union leaders to do better at the bargaining table. By requiring that employers publicize both their own lost-time injury rates and the industry average, OSHA could pressure unions to pay more attention to health and safety matters by exploiting the desire of unions to standardize terms and conditions of employment.

Expanding the NIOSH Health Hazard Evaluation Program
Section 20(6) of the Occupational Safety and Health Act gives NIOSH authority to evaluate whether a substance found in a workplace is toxic at prevalent levels of concentration. (Evaluations are initiated by a written request from either the employer or an "authorized representative of employees" that specifies "with reasonable particularity the grounds on which the request is made.") This service is known as the Health Hazard Evaluation Program. It is particularly valuable to unions in that it gives them a "cheap" means of assessing hazards that would otherwise be beyond their technical capability. The union does not have to bargain with management to obtain permission for an outside hazard evaluation as it would if the evaluation were to be performed by a consultant. Moreover, unlike an OSHA inspection, the NIOSH evaluation pro-

vides the union with workplace-specific information concerning unregulated hazards, thus facilitating negotiation over hazard-limiting agreements.

Notwithstanding the fact that it directly addresses some of the technical problems associated with health and safety bargaining, the Health Hazard Evaluation Program has been consistently underfunded and underpublicized since its inception. NIOSH performs only about 100 health-hazard evaluations annually, with a total budget for the program of around $2 million.[8] Because of the resource constraint, NIOSH has rationed health-hazard evaluations by rather strict construction of the statutory requirement that the reason for a requested evaluation be specified with particularity.[9] General requests that do not allege a causal relationship between some agent and an observed physiological response are typically denied. Additional funding would permit NIOSH to be more responsive to union requests for technical assistance. Properly targeted section 20(6) evaluations could strengthen unions' bargaining positions by providing independent assessment of the risks posed by unregulated hazards.

Expanding the Right to Refuse Hazardous Work

If unions' bargaining agendas were determined by the most safety-conscious workers, health and safety issues would receive higher priority. Government obviously cannot mandate this, but it can attempt to give more intraunion political power to the safety-conscious as a means of redressing their systematic underrepresentation. Expanding the statutory right of workers to refuse hazardous work would accomplish this objective. It also would strengthen the overall bargaining position of unions on health and safety issues and, of course, remove individuals from actual exposure.

The power of a union derives not only from its ability to withhold labor services from management, but also from its ability to guarantee the uninterrupted supply of labor during the term of its contract (usually by pledging not to strike[10]). Interruptions in the supply of labor despite a no-strike pledge threaten the authority of the incumbent union officers. If a strike is sanctioned by the union, management can seek injunctive relief and damages for breach of the terms of the collective bargaining agreement. Even if not sanctioned, such interruptions place pressure on the union hierarchy to address the issues that underlie the dispute. The pressure may come either from the organizers of the work stoppage, who pose a political threat to the incumbents, or from management, which seeks relief from production delays. Expanding the right of workers to refuse hazardous work gives more power to safety-conscious workers because it permits them to draw the attention of both the union and the management to their specific complaints. Even if the union is indifferent to the individual worker, management may seek resolution of the underlying issues in order to avoid future work stoppages. If individual workers acting on their own can legally refuse hazardous work without threat of job loss, safety-conscious workers can pressure their union to pay more attention to health and safety issues. It is important that this right not be collective; if it can only be exercised with the consent of the union it will have little impact on union bargaining priorities.

The legal dimensions of the right to refuse hazardous work are defined by statute, judicial decisions, and administrative rulings. The body of law has evolved with little consideration of the impact of this right on collective bargaining over health and safety issues. The protection it offers a worker depends on the legal authority for his claim.[11] Under the Labor Management Relations Act, workers can refuse any hazardous work if the refusal constitutes "concerted activity for . . . mutual aid or protection."[12] The key words are "concerted" and "mutual." A walkoff by an individual worker for his exclusive benefit is not protected because it is not concerted activity. Thus, the LMRA offers little opportunity for the individual safety-conscious worker to pressure his union or employer to address health and safety issues by refusing hazardous work. Moreover, walkoffs are not protected at all under the LMRA if the collective bargaining agreement contains a no-strike pledge.[13]

In theory, section 502 of the Taft-Hartley Act, which states

nor shall the quitting of labor by an employee or employees in good faith because of abnormally dangerous conditions . . . be deemed a strike under this chapter,

should offer the threatened worker some protection.[14] If a walkoff legally is not a strike in derogation of the contract, then the worker cannot be discharged for refusing the hazardous work. Moreover, since section 502 speaks of "an employee," arguably the protected right is personal and need not be asserted in concert or for the benefit of others. In practice, though, section 502 affords workers very little protection for two reasons. First, it covers only exceptional risks; a worker who regularly faces excessive risk of injury has no rights under section 502. And second, the Supreme Court has interpreted the "good faith" provision as requiring that the worker prove the existence of an abnormal danger by "ascertainable, objective evidence."[15] Thus, a safety-conscious worker who makes a reasonable but subjective determination that a danger is exceptional risks loss of pay or employment if his assessment proves erroneous.

The Supreme Court further constrained the right to refuse hazardous work in *Gateway Coal v. United Mine Workers*[16] by extending the judicial presumption of arbitrability to health and safety disputes. The court's decision was based on the belief that it is better to resolve such disputes through arbitration than through industrial strife.[17] Accordingly, the court held that since a refusal to work because of allegedly unsafe working conditions was an arbitrable dispute, the related work stoppage could be enjoined unless the right to refuse hazardous work was expressly excluded from the no-strike pledge.

The court's view that arbitration of health and safety disputes and a right to refuse hazardous work are mutually exclusive is misplaced. It is possible—indeed desirable—for workers to be able to refuse work they suspect is hazardous and then have their judgment tested by expedited arbitration. This is the procedure employed by some firms in Basic Steel, as described in chapter 5. Triggering the arbitration of safety disputes by refusing to work simultaneously removes workers from risk and creates symmetrical incentives for labor and management to resolve the underlying health and safety issue. These benefits are in addition to the incentives such a right creates for union attention to health and safety matters. (The proper role of arbitration of health and safety disputes is addressed below.)

In 1973 the Department of Labor attempted to expand the right to refuse hazardous work by ruling that there are implied as well as express rights articulated in the Occupational Safety and Health Act. Included in these implied rights is a limited right to refuse hazardous work when an employee concludes "that there is a real danger of death or serious injury and that there is insufficient time, due to the urgency of the situation, to eliminate the danger through resort to regular statutory enforcement channels." [18] Accordingly, the department has held that this implied right is protected by section 11(c) of the OSH Act, which prohibits an employer from discriminating against a worker who exercises "any right afforded by this chapter." Although the Supreme Court recently upheld the Labor Department's position in a unanimous decision,[19] it did not settle the issue of whether an employee is entitled to back pay for the time he did not work because of the life-threatening condition. The Court merely held that an employer could not punish an employee for refusing a hazardous assignment.[20] While the Court's decision protects such a worker from loss of job, it still does not go far enough. If the right to refuse hazardous work is to be meaningful, a worker must be entitled to back pay for the period in which he legitimately refused dangerous work.[21] A statutory clarification of this issue would provide the proper incentives to workers. Moreover, providing back pay to workers in such situations would give management an additional incentive to resolve health and safety disputes quickly. Most important, a strong right to refuse hazardous work would give a lot more clout to safety-conscious workers.

Promoting Swift Arbitration of Health and Safety Disputes

The availability of a fast, effective means to resolve health and safety disputes is important to the development of collective bargaining over job hazards for three major reasons. First, as noted in chapter 6, collective bargaining agreements are implemented through the grievance-arbitration process. To the extent that the parties do not have mutual confidence in this process, the implementation of hazard-limiting agreements will be inhibited. Second, arbitration acts to fill in the inevitable gaps in a collective bargaining agreement. Since it is virtually impossible to anticipate all health and safety contingencies in a single written agreement, arbitration permits resolution of future disputes without resort to economic conflict. And third, as suggested by the preceding section, the availability of arbitration clearly influences the significance of the right to refuse hazardous work. From labor's perspective swift arbitration protects the worker who rightfully refuses hazardous work against improper discipline, and from management's perspective it guards against abuse of the walkoff right.

Giving Arbitrators Health and Safety Training
Not all issues are amenable to arbitration. Labor and management contractually agree to submit disputes to arbitration only when they have confidence in the ability of the arbitrator to resolve disputes fairly and in a manner that accurately reflects the realities of the workplace. But, as discussed in chapter 6, health and safety disputes pose unusual problems for arbitrators, whose professional training is almost always based either in law or in labor relations. Non–health-and-safety grievances typically involve simple factual questions or legal interpretations of contractual language: Did the employee report for work drunk? Does the contract permit "contracting out" of maintenance work? Though the answers to these questions may be unclear, at least the complexities are in provinces in which an arbitrator possesses expertise—labor relations and statutory construction. In contrast, the complexities of health and safety disputes are rarely well understood by arbitrators. Not only are most arbitrators technically ignorant; they also lack the practical knowledge of specific hazards that comes from working on a particular job for a number of years. Thus, under present circumstances, arbitrators often are less well equipped to resolve health and safety disputes than the parties themselves. This situation is not immutable. OSHA can try to upgrade the quality of health and safety arbitration by funding a pilot training program for arbitrators in hazardous industries. Specifically, a training program for arbitrators might cover relevant OSHA standards, conditions that pose extreme danger to workers, the types of evidence commonly offered to establish the existence of exceptional risks, and other technical subjects that would assist in the resolution of disputes related to negotiated health and safety agreements.[22]

Paying for Health and Safety Arbitration

Training arbitrators will not ensure arbitration of health and safety disputes. Parties arbitrate grievances only if they elect to do so contractually. To the extent that other dispute-resolution mechanisms are cheaper, the parties will not negotiate arbitration clauses. As noted earlier, the OSHA complaint procedure creates disincentives for arbitration of health and safety disputes because it is a free resource to unions whereas arbitration is not. By underwriting the cost of arbitration of health and safety grievances, OSHA could redress this imbalance. If arbitration is a substitute for OSHA complaints, as posited, then funding the cost of the arbitrator should be relatively inexpensive because it would reduce the demand for OSHA's services.

If these efficiency gains are to be realized, workers should be barred from pursuing an OSHA complaint after adjudicating a grievance through arbitration. Furthermore, if OSHA funding of arbitration is to succeed, arbitrators must enjoy the same degree of independence as under the current system of arbitration. The parties must be free to select the arbitrator of their choice; OSHA should not certify arbitrators for health and safety arbitration, but should merely agree to contribute to the cost of the arbitrator's salary if the arbitrator has completed an OSHA-approved health and safety course. OSHA should rely on the parties to screen the arbitrators for competence.

OSHA should encourage arbitration of health and safety disputes not because it is a substitute for government regulation, but because it facilitates implementation of negotiated health and safety agreements. Even with the training suggested above, arbitrators will not be able to enforce general health and safety provisions that merely require that "the company shall make reasonable provision for the safety and health of its employees at its plants during the hours of their employment." If OSHA and NIOSH are incapable of specifying what constitutes a reasonable exposure to a known carcinogen, then an arbitrator will not be able to do so either. Arbitrators can, however, resolve disputes concerning interpretation of more specific provisions of negotiated agreements. For example, an arbitrator can determine whether foremen are complying with a steel company's pledge to stage-charge its coke ovens, whether oven doors are being repaired according to the schedule specified in a collective bargaining agreement, and whether it was reasonable for an employee to refuse to perform work he believed abnormally dangerous. Arbitration contributes to an overall strategy for encouraging negotiation of health and safety agreements because it helps labor and management live and work with the rules they define at the bargaining table.

Promoting Union Participation in Health and Safety Education Programs for Workers

A significant proportion of job accidents are caused by factors directly related to day-to-day work practices and habits. In theory, training programs are the only means available for influencing these accidents. Also, where outside inspection is of limited usefulness in enforcing regulations because of the transient character of the job site, training workers to spot standards violations may motivate them to pressure management to achieve compliance. However, as noted in chapter 6, most unions lack the expertise necessary to develop effective health and safety training programs. Some unions also lack the inclination.

Training and education programs are cheaper than inspection programs, they engender relatively little political opposition, and (in contrast with other union health and safety initiatives) they often can be instituted without the contractual assent of management. To maximize the effectiveness and the efficiency of its training efforts, OSHA should concentrate on integrating health and safety education into existing union apprenticeship programs by offering to each union technical assistance that is geared directly to the work practices and the hazards encountered by members of the union. In effect, OSHA must separately assist each union in designing an individual health and safety education program that fits neatly into the existing apprentice training program. Moreover, OSHA must be prepared to market its health and safety consultation services aggressively. If the agency waits for unions to seek its assistance, many apprentice-training curricula will go unchanged.

Channeling training expenditures into apprenticeship programs is likely to be more effective than establishing remedial training programs for four main reasons. First, presumably it is easier to teach safe work habits initially than to break acquired bad habits. Moreover, to the extent that inexperienced workers suffer a disproportionately high number of accidents, concentrating training in apprenticeship programs addresses the most severe aspect of the problem. Second, once apprenticeship curricula have been modified to incorporate health and safety material, few if any additional expenditures are needed to maintain the program. Third, workers treat apprentice training more seriously than "extracurricular" education activities because their ability to pursue their occupation is usually conditioned upon successful completion of the program. And fourth, directing health and safety training through apprentice programs provides another way of involving unions in health and safety matters generally.

Little is known about the impact of health and safety training on the rate of occupational accidents. While it seems reasonable that such training should inspire safer work practices (especially if it has been part of a worker's professional training), little empirical evidence is available. In funding training programs, OSHA should try to gather information on their effectiveness by requiring evaluations that focus on the impact of training on the lost-time injury rate. Specifically, OSHA should require training contractors to perform longitudinal studies that compare the accident rate of locals that receive health and safety training with that of locals that do not. The training programs presently funded are not evaluated in terms of outcome measures. Instead, OSHA relies upon tests, questionnaires, and interviews to determine effectiveness. If contractors are incapable of obtaining the data necessary to evaluate outcomes, OSHA should track the accident histories of locals or firms that have instituted health and safety training programs. In light of the large proportion of accidents that are unabatable by any means other than training, and the minimal compliance incentives created by inspections, a larger commitment to worker training and education appears justified even in the absence of data on effectiveness.

Conclusions

If OSHA is to be even modestly successful in controlling hazards in the workplace, it cannot "go it alone." The job of identifying and abating occupational hazards is beyond the capabilities of a single federal agency. OSHA badly needs allies, and the most logical place for it to turn is the constituency it is supposed to protect: workers. Unions can do lots of things that OSHA cannot, from identifying momentary hazards to pressuring line supervisors to comply with health regulations. Moreover, with a little help, they can perform these tasks cheaply and effectively. If OSHA is creative in its use of incentives, it can enlist the help of organized labor in the pursuit of regulatory objectives.

The suggestions for reform in this chapter are not radical departures from present practice. Rather, they represent modest attempts to improve regulatory performance through small changes in the way OSHA conducts its business. Most of these changes can be implemented without new legislation. Also, most are not likely to provoke substantial political opposition. They should be tried on an experimental basis and evaluated. If successful, they should be implemented on a larger scale. If not, they should be scrapped and the agency should look for other ways to make the workplace safe. What is important in this process of experiment and evaluation is that OSHA not lose sight of the inherent characteristics of the occupational safety and health problem that make successful regulation so difficult. Difficult problems demand innovative solutions. If OSHA continues to be bound by the existing regulatory structure, it will fail.

8
Lessons for the Design of Regulatory Policy

What does our collective-bargaining-based approach to the design of job safety and health policy tell us about other types of regulatory problems? After all, collective bargaining will not control hospital costs, make the nation's air cleaner, or render buildings safe for habitation. What is it about collective bargaining as a regulatory strategy that is generalizable to other regulatory problems?

Diversity of problems, which complicates the design of occupational safety and health policy, also complicates the design of other types of regulatory policies. Also, the way we now regulate job hazards—through standards and fines—is not all that different from the way we regulate lots of other problems. The reform measures described in the preceding chapter address the problems created by the diversity of conditions by capitalizing on the self-interest of the protected constituency (labor) to achieve flexibility within the "command and control" regulatory process. This approach has applications to other regulatory problems. Furthermore, collective bargaining increases the likelihood of effective implementation of standards by using a consensual approach to resolving differences over regulatory compliance. Finally, perhaps the most important lesson that emerges from the analysis of policy toward job hazards is that different aspects of regulatory problems frequently demand different policy solutions.

Tailoring Policies by Flexible Enforcement

Because of the diverse nature of the world, regulatory policies must vary across a number of dimensions if they are to be effective. A good policy must take into account the facts that different problems demand different solutions, that problems vary over time and over geographic regions, that the effectiveness of different solutions depends on the regulated industry's market structure and the competitive positions of firms within the industry, and that the internal organization of the regulated enterprise greatly influences the prospects for successful implementation of regulatory policy.

In practice, it is difficult for a regulatory agency to structure its policies to accommodate these differences for several reasons. Agencies lack the rulemaking resources necessary to form precisely appropriate "command and control" policies. Knowledge of special conditions is always in short supply, as is the personnel needed to write extremely detailed regulations. Even if an agency were capable of writing regulations that reflected the diverse conditions encountered in the world, the regulations would be so complex as to be incomprehensible. The Internal Revenue Code and its accompanying regulations are a good example of the confusion that reigns when government tries to be specific. Also, appearances matter—policies that draw rational but subtle distinctions among regions, industries, and firms often appear unfair to those who lack the time or capacity to review the supporting analysis. And truly decentralized regulatory strategies, such as taxes, liability rules, and economic incentives, have proved politically unpopular.

In a world constrained by political and administrative realities, it may not be possible to deal with the diversity of conditions either through a decentralized policy or through a direct policy that permits multiple exceptions. Blunt "command and control" regulations are the rule rather than the exception, and although less centralized regulatory strategies might be preferred there is little chance that Congress will scrap OSHA, TSCA, and other centralized regulatory schemes in the near future. Thus, at least in the short run, if we wish to capture some of the efficiency gains that might be realized through decentralized strategies, we may have to do so within the context of centralized "command and control" regulation. The strategy described in chapter 7—delegating some enforcement responsibility to the protected constituency—includes a measure of decentralization (and efficiency) through flexibility that is introduced through the implementation process.

Critics of "command and control" regulation often assume that standards are implemented as rigidly as they are defined. But implementation need not be rigid, and in many cases is not. Enforcement processes often "adapt on their own" when absolute enforcement is inefficient or impossible. Even policies that appear to prohibit absolutely the balancing of costs and benefits are often unsuccessful in blotting out efficiency concerns. For example, the Endangered Species Act forbids development that threatens the critical habitat of any species listed by the Secretary of the Interior as endangered. Consideration of endangered status is supposed to be a technical decision. The act does not permit comparison of the benefits forgone from development with the value of the species to society. But in practice, not all endangered species have been listed and not all development curtailed. Listing decisions and critical-habitat designation have often been made in ways that minimize the adverse impact on development.[1] To be sure, the implementation of the Endangered Species Act cannot be characterized as economically efficient, but it is certainly more efficient than it is given credit for. Similarly, although the Delaney Amendment to the Food and Drug Act calls for the banning of any substance known to be carcinogenic, the Food and Drug Administration has found ways to avoid banning coffee, tea, selenium, pepper, and peanuts.

There is a good reason why implementation processes occasionally reflect efficiency considerations: It is difficult to ignore them. An agency that attempts to implement a regulation that is obviously ineffective or that imposes extraordinary costs upon regulatees exposes itself to public scorn and ridicule. Furthermore, it is very difficult for a regulating agency to get away with claiming that it will ignore costs in assessing regulatory alternatives. Inevitably, someone will come along and do the cost-benefit calculation the agency refuses to do. Soon after OSHA claimed that costs were not a relevant consideration in setting the proposed coke-oven standard, an article in *The Public Interest* pointed out that compliance would cost between $8,000 and $43,000 per exposed worker per year.[2] In practice, public debate over alternative policies is conducted in a way that suggests that there is a "law of conservation of cost-benefit analysis" at work: Given any regulatory proposal, some group will be interested in estimating the likely costs and benefits.[3]

The collective-bargaining-based regulatory strategy described in chapter 7 represents an attempt to enhance the capacity of the implementation process to reflect efficiency considerations. In effect, it constitutes decentralization through an enforcement process that is flexible by design. Rather than attempt to enforce regulations centrally, it capitalizes on the self-interest of the protected constituency to achieve compliance. It gives responsibility to unions for enforcing regulations while at the same time building their capacity to do so. Because unions have better knowledge than OSHA of local conditions, local preferences, and local implementation problems, the resulting pattern of compliance should be both

Lessons for the Design of Policy

more efficient and more effective than what might be realized from a typical "command and control" strategy. It will not be as efficient as the decentralized intervention strategies urged by many economists, but it will be more efficient than what we have now.

A fair number of opportunities could probably be found to improve the effectiveness and efficiency of regulation through innovative enforcement procedures if we would only look for them. In general, where the beneficiaries of regulation are identifiable, ceding to them some enforcement authority (as the collective bargaining strategy does) should yield benefits similar to those offered by decentralized intervention strategies. For example, giving local communities responsibility for enforcing federal outdoor noise standards should bring about compliance more efficiently than if the federal government enforced the standards uniformly. If local government is representative and responsive, it will enforce the noise regulations in a manner that is consonant with local interests;[4] some regulations will get enforced as written while others will be amended through the implementation process when strict compliance proves too costly for the community or the regulatee or both.[5] Thus, delegation of enforcement responsibility introduces the desired flexibility into the enforcement process. Delegation will not work if the beneficiaries of regulation are diffuse or poorly organized. Similarly, if the interests of some of the beneficiaries are in opposition it may not be practical to delegate enforcement responsibilities to a single entity. For example, it is not practical to entrust enforcement of water pollution regulations to individual municipalities because of the externalities associated with pollution-abatement activities, although county or regional government might be able to discharge such duties effectively. When delegation is impractical, expanding the boundaries within which compliance must be achieved will also bring about gains in regulatory efficiency. The U.S. Environmental Protection Agency has recently begun to experiment with this approach. Under its "bubble policy," instead of setting emissions standards for each individual operation within a plant the EPA sets one overall standard governing emissions from the entire activity. Because the regulatee is free to make tradeoffs within the compliance "bubble," the policy should produce a more cost-effective pattern of compliance.

Innovation in enforcement policies may bring about many of the same improvements in the efficiency of regulation that might be expected from decentralized intervention strategies. In many cases, such improvements can be achieved within existing regulatory frameworks. Identifying opportunities for making the enforcement process flexible and more efficient will require detailed attention to the minutiae of implementation. Although the policies that result from such an inquiry will be a long way from being economically efficient, they almost certainly will be an improvement over current practice.

Consensual Approaches to Regulatory Decisionmaking

Regulation today is in large part a legal process. Regulatory authorities rely upon legal proceedings to enforce regulatory obligations, and regulatees rely upon the courts to challenge the legal authority of the regulators. In many cases, the availability of subsequent legal remedies greatly influences both the form and the substance of regulatory policymaking. Regulatory proceedings more often than not resemble trials, with administrative law judges presiding and with opposing counsel objecting to the introduction of evidence and cross-examining witnesses. Hearings often become opportunities to jockey for position in the litigation that inevitably follows major regulatory decisions. The nature of this process frequently drives a wedge between opposing interests as each side adopts the most extreme defensible position before the agency. As with trials, this process invariably produces winners and losers. In those cases where the agency adopts a middle position, it is not unheard of for both sides to attack the decision.

Enforcement and permit-granting actions are not very different. In many cases, such proceedings are not over until the last available court has refused to hear the appeal of the last litigant claiming that the agency action was either too harsh or too lenient. Judicial affirmation of agency action does not necessarily ensure successful implementation, since courts are rarely in a position to supervise or influence the day-to-day activities of regulatees.

With respect to enforcement activities, at least, the collective bargaining strategy described in chapter 7 represents a fundamen-

Lessons for the Design of Policy

tally different approach to regulatory decisionmaking. It consciously attempts to bring together regulatees and the beneficiaries of regulation to settle their differences without resort to either the courts or an outside agency. This consensual approach offers a number of advantages over the litigation model. As noted in chapter 5, in the case of health and safety issues nonadjudicatory dispute resolution is faster. Also, the outcome of an enforcement process that relies upon bargaining to settle disputes reflects the preferences of the parties instead of those of the enforcing authority. Further, bargaining processes do not produce clear winners and losers. (For example, contrast the ritual that labor and management go through upon completion of contract negotiations, as each side claims that the agreement is good, with the postlitigation comments of plaintiffs and defendants.) In addition, as the parties responsible for implementing an agreement bargain over its terms, they can bind themselves to undertake actions required for successful implementation.

With these obvious advantages, can the concept of bargaining be extended to forms of regulatory decisionmaking beyond enforcement of job safety and health regulations? The answer is a qualified yes. There are more and more attempts to apply conflict-resolution techniques developed in the course of labor-management negotiations to environmental disputes.[6] When John Dunlop was Secretary of Labor he tried (unsuccessfully) to bring labor and management together to negotiate the terms of what would become the OSHA coke-oven standard. The National Coal Policy Project brought together representatives of industry and the environmental movement to negotiate over the dimensions of national coal policy. Each of these attempts to integrate bargaining into the regulatory process labored under a number of handicaps that do not arise in the context of enforcing job safety and health rules. These issues must be addressed if bargaining is to play a more significant role in general regulatory decisionmaking.

Identifying the Parties
It is clear what parties should be involved in negotiations over compliance with safety regulations in a particular plant: labor and management. The parties are well organized and their interests

well defined. But it is not obvious who should be party to the negotiations over whether a dam should be constructed in a wildlife area. Who, for example, speaks for the environmental interests? Should each of the communities downstream be involved? Because the legitimacy of bargaining as a dispute-resolution procedure rests on the participation of the interested parties, procedural mechanisms must be developed for ensuring that all legitimate interests are represented in any negotiations that might lead to a consensual resolution of a regulatory dispute.

Creating an Environment Conducive to Negotiation

A common characteristic of negotiations is that when bargaining becomes serious it inevitably becomes private. Most observers of labor negotiations interpret a news blackout as a sign that a contract is near, and international negotiations are usually conducted in complete secrecy. There is a good reason why the public is always excluded from serious bargaining sessions: The give-and-take that is the essence of successful bargaining is inhibited if the parties must be concerned with how their constituencies will interpret intermediate positions taken on specific issues. Since it is not possible to win on every issue, negotiators prefer to present the fruits of their labor as a package instead of piecemeal. Secrecy permits this.

Consequently, if regulatory policy is to be determined through a process of negotiation, then we must be prepared to tolerate some secrecy. But this runs counter to the current trend in the United States to give the public more access to policymaking processes, not less. For example, recently enacted "sunshine laws" require that either the public be admitted to most government meetings or that transcripts be made available shortly after most closed meetings. Similarly, the Administrative Procedures Act severely limits *ex parte* communications between regulators and affected interests. These laws, which were intended to strip away the shroud of secrecy covering most government decisions, also have made it impossible for concerned parties to engage in the informal give-and-take that is essential to successful bargaining. If we are to create an environment conducive to the negotiation of consensual agreements, then we must carve out exceptions in the above laws to permit bargaining to proceed in private.

Providing Incentives to Negotiate in Good Faith

Bargaining over job safety and health issues offers labor and management the chance to strike deals that leave both sides better off. Consequently, both labor and management have an incentive to negotiate in good faith and seek agreement. However, because of the way alternatives are presented in other regulatory contexts, bargaining is often of the sort where what one party gains the other loses. One side may refuse to negotiate or may adopt an obstructionist position if it prefers the status quo to any alternative under consideration. For example, if a state energy-facility siting board is considering licensing a new recycling center, neighbors of the proposed facility may perceive that it would foul their air, pollute their water, and congest their streets. If the decision to license the facility is presented as a yes-or-no choice, the neighbors will be obstructionist because construction of the facility will leave them unambiguously worse off. The neighbors will relax their position only if alternatives are considered that might improve their net welfare. For example, if the developer were to offer to compensate the neighbors for the diminution in property values caused by the facility, the neighbors might bargain constructively because they might prefer the package of the facility plus compensation to no facility at all. The possibility of compensation payments changes the character of the bargaining and in the process provides the parties with incentives to negotiate in good faith.[7] If we can package regulatory alternatives in a way that encourages this type of bargaining, we can expand significantly the number of regulatory disputes that might be resolved through consensual agreement.

Binding Parties to Agreements

Labor-management agreements are relatively easy to enforce. Not only does each side have the legal authority to bind itself, but unions also have the power to bind their individual members. If the union leaders assure management that a dispute is settled, the dispute usually is over. If the rank and file disagrees with the terms of the settlement, its only legal alternative is to vote the union leaders out of office at the next election. A second reason labor-management agreements are easily enforced is that the continuing relationship between the parties encourages fulfillment of contractual obligations. Each side has leverage against the other to enforce the agreement's terms. Furthermore, the fact that each side will face the other in future contract negotiations also creates pressure for compliance.

It is not so easy to bind parties to agreements in other disputes involving large organizations with diffuse memberships, or in disputes in which local governments are parties. Although an organization such as the Sierra Club can legally bind itself to abide by an agreement, it cannot also bind its individual members. Consequently, while the leadership of a large group may agree as part of a consensual agreement not to challenge a regulatory decision in court, the agreement is not binding on individual members, who legally may resign their membership, form a new organization, and sue to enjoin the regulatory decision.[8]

Local governments have a difficult time binding themselves for an entirely different reason. It is a well-settled legal principle that governments cannot contract away their policymaking powers. For example, most state courts would refuse to enforce a contract between a developer and a community in which the community agreed to rezone a site to permit construction of a facility in return for a promise by the developer to compensate neighboring residents. Although the community could consider the existence of the compensation agreements in deciding whether to rezone, a contract that mandated rezoning would be void. Thus, there are substantial legal limitations on the abilities of diffuse interest groups and local governments to offer commitments to opposing interests that would ensure compliance with a negotiated agreement. Moreover (at least in the environmental field, where disputes often center around the impact of a new activity on the local environment), the parties usually do not have the same type of continuing relationship that encourages compliance in the labor-management arena. In practice, this inability to ensure compliance may inhibit efforts to settle regulatory disputes through consensual agreements. If such agreements are to become common, legislation may be needed to facilitate the binding of parties.

Matching Regulatory Tools to Regulatory Problems

The regulatory tactic articulated in chapter 7 addresses only part of the occupational safety and health problem. It would improve conditions for workers in unionized workplaces. However, aside from the fact that it would free some inspectors to look elsewhere, it would do little for nonunionized workers. The limited scope of this approach is not a major failing. What this strategy does is use different policy tools to address different aspects of a particular regulatory problem. That an innovative policy addresses only part of a problem should not lead us to reject it, but rather should inspire us to look for other ways to deal with the rest of the problem.

Academics have a general tendency to look for global solutions to interesting policy problems. In fact, much of the academic debate over regulatory reform centers on the choice between standards and economic incentives. Although the participants in this debate have produced much interesting literature, the debate itself has been over the wrong question. Economic incentives can no more be preferred to standards than a wrench can be preferred to a pair of pliers—there is a time and a place for each, and what is important is knowing when to use which. To do a good job, the regulator needs to know how to match regulatory tools to regulatory problems.

The academic literature has not had much to say on this important subject because it has focused almost exclusively on one criterion—economic efficiency—in determining congruence between tools and problems. To be sure, if efficiency is the only criterion economic incentives will almost always be preferred to other regulatory instruments. But regulators have to worry about things other than efficiency. They have to be able to implement and manage programs effectively, and their ability to implement and manage is determined by inherent characteristics of the problems they are trying to address. For example, standards may be preferred to taxes in situations where it is difficult to continuously monitor the taxable event, and economic incentives might be preferred where the number and the locations of regulatees makes inspection to determine compliance with standards infeasible. If we are to develop a useful theory for matching tools to problems, then the criteria used for evaluating the match must reflect not only efficiency considerations, but also the managerial, institutional, and political factors that determine the effectiveness of policies in practice.

In addition to considering additional criteria in evaluating regulatory alternatives, we need to give more thought to how we define regulatory problems. The description of a problem affects our perception of what constitutes an acceptable solution. For example, to the extent that we consider occupational disability to be one regulatory problem, we are likely to look for one policy that responds to all of the problem. In practice, lots of factors determine how a problem gets bounded or described, including the jurisdictional mandate of the institution that is responsible for addressing the problem, the professional norms of the people who diagnose the problem, and the structure and organization of the constituency that calls attention to the problem. Since there are lots of ways to bound a problem, the boundaries in use cannot be considered sacred. In some cases, proposed solutions will suggest problem boundaries that not only are respectable, but also assist us in matching tools to problems. For example, the case for treating unionized and nonunionized workplaces differently rests on the recognition that the existence of a union creates an opportunity to supplement health and safety regulation that does not exist otherwise. In general, regulatory policy would be both more efficient and more effective if we identified and exploited modest but significant opportunities to do better. And in many cases the only way we will succeed in identifying these opportunities is by asking the simple question "What works?"

Notes

Chapter 1

1
President's Report on Occupational Safety and Health, 1971, vol. 11, no. 7, p. 30.

2
National Safety Council, *Accident Facts* (1975).

3
R. S. Smith, *The Occupational Safety and Health Act* (Washington, D.C.: American Enterprise Institute, 1976), p. 5.

4
Ibid., p. 7.

5
U.S. Department of Commerce, *Historical Statistics of the United States: Colonial Times to 1970* (1976), p. 77.

6
A more cynical observer of regulatory politics might argue that opponents are reluctant to raise the issue of policy efficacy for fear that the resultant debate might produce a more effective policy.

Chapter 2

1
Occupational Safety and Health Act of 1970, section 6(a).

2
Executive order 12004 (*Federal Register* 41 [March 23, 1978]: 12661) requires preparation of a Regulatory Analysis for proposed regulations that have major economic consequences. The analysis must contain a succinct statement of the problem, a description of regulatory alternatives, and an analysis of the economic consequences of each alternative.

3
Marshall v. Barlow's Inc., 46 L.W. 4483, 429 U.S. 1347 (1978).

4
Probable cause can be established by reference to an employee complaint, by the prior accident or health record of the firm to be inspected, or by establishing that the firm was selected on the basis of a general plan that utilized neutral sources of information to target inspections. (Ibid. at 4487.)

5
The OSH Act provides that individual states may assume responsibility for regulating job hazards provided that the state regulatory plans are at least

as effective as the federal program. State plans must be approved by the Secretary of Labor.

6
A. Nichols and R. Zeckhauser, "Government Comes to the Workplace: An Assessment of OSHA," *Public Interest* 49 (1977): 39–69.

7
U.S. Senate, Committee on Labor and Public Welfare, Subcommittee on Labor, OSHA Review 1974, Appendix II, Issue Papers, p. 967.

8
J. Mendeloff, *Regulating Safety: An Economic and Political Analysis of Occupational Safety and Health Policy* (Cambridge, Mass.: MIT Press, 1979), p. 156.

9
J. Q. Wilson, "The Politics of Regulation," in J. McKie, ed., *Social Responsibility and the Business Predicament* (Washington, D.C.: Brookings Institution, 1974), pp. 151–152.

Chapter 3

1
J. Mendeloff, "An Evaluation of the OSHA Program's Effect on Workplace Injury Rates: Evidence from California Through 1974," U.S. Department of Labor, contract B-9-M-5-2399 (1976).

2
Ibid., p. 61.

3
R. Smith, "The Estimated Impact on Injuries of OSHA's Target Industry Program," Cornell University (undated).

4
A. Di Pietro, "An Analysis of the OSHA Inspection Program in Manufacturing Industries 1972–1973," draft technical analysis paper, Office of the Assistant Secretary for Policy, Evaluation, and Research, U.S. Department of Labor (1976).

5
W. K. Viscusi, "The Impact of Occupational Safety and Health Regulation," *Bell Journal of Economics* 10 (1979): 117–140. Only 22 industries are included in the 1972 sample.

6
U.S. Bureau of Labor Statistics news release, December 8, 1976, p. 5.

7
Job Safety and Health 4 (May 1976): 3.

Notes to pp. 29–40

8
U.S. Department of Labor, "Inflationary Impact Statement: Coke Oven Emissions" (undated), p. 81.

9
McGraw-Hill, *Third Annual Survey of Investment in Employee Safety and Health*, cited in Mendeloff, "An Evaluation of the OSHA Program's Effect on Workplace Injury Rates," table B-1.

10
Cited in D. C. Kallis, "Update on the Cost of OSHA Compliance," *Occupational Hazards* (August 1974): 42.

11
U.S. Department of Labor, "OSHA Inflationary Impact Statements" (undated).

12
R. Zeckhauser and A. Nichols, "The Occupational Safety and Health Administration: An Overview Prepared for the Senate Committee on Government Operations" (95th Congress, first session), p. 52.

13
No mention is made at this point of the criteria that should be used to make such a calculation; the model is intended to be non-normative. Later sections will evaluate both the current regulatory approach and alternative intervention strategies to determine how well each makes the types of tradeoffs implied above.

14
In addition to requiring compliance with standards, section 5(a) of the OSH Act also imposes a general duty on an employer "to furnish to each of his employees employment and a place of employment which are free from recognized hazards that are causing or likely to cause death or serious harm to his employees." While this "general duty clause" imposes a legal duty on employers to abate hazards not covered by standards, it offers no assistance to the employer in identifying the dangerous conditions that require abatement. All this section of the act does is marginally supplement the financial incentives provided by the marketplace to identify hazards.

15
M. Corn, "Status Report on OSHA," U.S. Department of Labor, January 12, 1977, p. 24.

16
Mendeloff, "An Evaluation . . . ," pp. 13–14.

17
Cited in W. Oi, "On Evaluating the Effectiveness of the OSHA Inspection Program," U.S. Department of Labor, contract L-72-86 (1975), p. 42.

18
If all other regulatory agencies adopt the same position, corporate profits will be treated as a "commons" and may then be overused or exhausted by regulators.

19
In the explanation of the final coke-oven standard, OSHA stated that it "does not believe it is appropriate to quantify even a range of the benefits of the final rule" because of the uncertainties involved in making such estimates. See *Federal Register* 41 (October 22, 1976): 4670.

20
Critics of this strategy often note that it will cause workers at different plants to be exposed to different levels of risk. There are two responses to this criticism. First, people who advocate uniform risk levels should recognize that uniformity can be achieved only at a cost that is measured not in dollars but in morbidity and mortality. For the same expenditure in hazard abatement, more lives can be saved if a little nonuniformity is tolerated. Second, unless risks can be standardized across all jobs, workers will sort themselves out so that people who prefer jobs that are riskier but are more interesting or pay more will be in riskier occupations. In effect, self-selection diminishes the apparent inequality of nonuniform risks.

21
The alternative to a specification standard is a performance standard that merely mandates a level of performance and leaves the determination of the method of hazard abatement to the firm (e.g., "noise levels shall not exceed 90 dBA"). If it were extremely costly to monitor performance, specification standards might be preferred to performance standards. This would occur if the efficiency gains from least-cost compliance with the performance standards were exceeded by the costs of measurement performance. This situation might arise if performance could only be monitored through destructive testing.

22
The duty to comply with OSHA regulations is owed by the employer to the government. Accordingly, it can only be waived by the government and not by the employees.

23
The GM-UAW memorandum of understanding on noise abatement indicates the difficulty of specifying a single noise standard for just one corporation: ". . . it is evident that the problem of noise varies in kind and intensity in each plant. Thus it is not feasible to establish a specific noise abatement program generally applicable throughout all the Corporation's facilities." The agreement left the design of noise-abatement programs to each local plant health and safety committee.

24
Other problems would still have to be ironed out. For example, does the worker retain his seniority if transferred to a new department? Is he entitled to his old rate of pay on the new job? What are the rights of a worker "bumped" by an OSHA-transferred worker?

25
An enforcement strategy that attempted to make an example out of a few firms would also probably create a market for bribes and side payments to inspectors.

26
Employees at all levels are immune from direct OSHA sanction. In theory, OSHA fines should provide incentives for all employees to implement OSHA regulations, since employers should alter existing employee incentive schemes to reflect the cost of noncompliance. In practice, relatively few such incentives trickle down because of the institutional insensitivity of existing incentive schemes to change.

Chapter 4

1
Zeckhauser and Nichols, p. 236.

2
R. Smith, "The Feasibility of an 'Injury Tax' Approach to Occupational Safety," *Law and Contemporary Problems* 38 (1974): 730–744.

3
Zeckhauser and Nichols, p. 223.

4
Mendeloff, *Regulating Safety*, pp. 168–169.

5
Ibid.

6
L. Bacow, "Regulating Occupational Hazards Through Collective Bargaining," National Technical Information Service report ASPER/PUR-77/1914/A (Washington, D.C., 1978).

7
It is in the employer's self-interest to consider his workers' preferences in addition to his own costs. The return on a particular investment in hazard abatement is the sum of forgone accident costs and reduced compensating wage payments. Worker preferences determine the level of compensating wage demands. Therefore, to maximize the return on hazard abate-

ment an employer must consider the preferences of his employees. Note that if all markets work perfectly, the employer will invest in hazard abatement until the last dollar expended returns just $1 in forgone accident costs and reduced compensatory wage payments. This will be the socially optimal level of hazard abatement. See W. Oi, "On the Economics of Industrial Safety," *Law and Contemporary Problems* 38 (1974): 669–699.

8
Zeckhauser and Nichols, p. 230; Bacow, p. 186.

9
It is necessary to tax exposure levels rather than cases of occupational disease for a number of reasons. First, because the onset of an occupational disease is not a distinct event, it is often not clear when the tax should be imposed. Second, many occupational diseases may be contracted as the result of cumulative exposure to hazardous substances in a number of successive jobs; taxing cases of disease only penalizes the last employer. Finally, a disease tax may be inefficient because some job-related diseases also have other causes. Lung cancer, for example, may be caused by smoking as well as asbestos dust.

10
Job Safety and Health 5 (December 1977): 24.

11
A case can be made that few organizations are structured to respond quickly to any changes in their operating environments. A look at the financial section of any major newspaper will reveal reports of bankruptcies of firms that failed to respond to changes in such fundamental matters as the demand for their product, the price of primary inputs, the nature of production technology, or the competitive structure of their industry.

12
The existence of workmen's compensation usually acts as a complete bar to all other forms of recovery in tort by the worker against the employer.

13
Employers with sufficient assets may elect to self-insure in some states.

14
R. Zeckhauser, "Medical Insurance, A Case Study of the Tradeoffs Between Risk Spreading and Appropriate Incentives," *Journal of Economic Theory* 2 (1970): 10–26.

15
D. Bok and J. Dunlop, *Labor and the American Community* (New York: Simon and Schuster, 1970), p. 208.

Notes to pp. 60–64

Chapter 5

1
Even if reliable outcome statistics could be obtained, other problems might frustrate the analysis. For example, it might be difficult to determine the direction of causation in a cross-sectional analysis. If more bargaining activity occurs where health and safety conditions are bad, bargaining may appear counterproductive. Similarly, a longitudinal study may make it difficult to separate the influence of OSHA and collective bargaining on employer hazard abatement.

2
As part of "pattern bargaining," which has existed in the auto industry for years, similar agreements were also negotiated with Ford and Chrysler. The representatives at Chrysler, however, were not made full-time until the 1976 contract. In addition, the minimum number of employees required to fund a full-time representative is larger at both Ford and Chrysler. GM plants with more than 10,000 employees get more than one representative. The shop committee chairmen may receive the health and safety training in plants with fewer than 600 employees.

3
One committeeman is elected for each 250 employees—the number that constitutes a district, except in smaller plants, which are allotted extra committeemen. Plants with 500 or fewer employees have three committeemen. Each committeeman is assigned a geographic zone within the plant for the purpose of dealing with grievances.

4
Shop committee members may also be district committeemen. The shop committee is the highest union governing body in the plant.

5
Personal communication, James M. Rhadigan, GM umpire staff.

6
The union health and safety representative holds his job for "an indefinite term." In practice, this means for life unless he neglects his duties or decides to abandon the job to pursue other union office. The purpose of the life tenure is to discourage reps from participating in union politics.

7
This difference is explained by a number of factors. Management is allotted a shorter period of time to respond to health and safety complaints at each step of the process, the entire shop committee does not have to be convened to consider a health and safety complaint, and briefs do not have to

be prepared by both sides in preparation for possible arbitration of the dispute.

8

The UAW can legally strike over two other issues during the term of a contract: speedup of production and rate of pay. Between 40 and 50 strike letters have been issued during the last 4 years in which health and safety issues were involved along with disputes over speedup or rate of pay. A few strikes have occurred over these joint disputes. It is impossible to say, however, whether these strikes were primarily motivated by health and safety concerns or other issues.

9

The 106 full-time health and safety representatives each earn approximately $15,000 per year. This understates the cost of the program somewhat because it does not include fringe benefits or training costs.

10

Source: *Federal Register* (October 22, 1976): 46746. In addition, coke-oven workers exhibit an excess incidence of nonmalignant respiratory diseases such as bronchitis and emphysema. Finally, coke-oven employees are exposed to the occupational hazards typically found in working environments that involve fire, high temperatures, and moving equipment.

11

To ensure a smokeless charge (known as stage charging), coal must be properly distributed among the four hoppers of the larry car. Sixty-seven percent of the charge must be contained in hoppers 1 and 4, 22 percent in hopper 2, and 11 percent in hopper 3.

12

Personal communication, James English, counsel for the United Steelworkers.

13

It is at least plausible that uncertainty concerning the availability of arbitration has actually contributed to the settlement of disputes short of arbitration. The risk of an adverse decision on arbitrability (which would be rendered by an arbitrator under current practice) may create incentives for both sides to settle.

14

Personal communication, James English.

15

See *Brough v. United Steelworkers of America, AFL-CIO*, 437 F. 2d 748 (1st Cir. 1971) and *Rawson v. United Steelworkers of America*, No. 12694 (Idaho Supreme Court, No. 12231).

Notes to pp. 73–75

16
To a large extent, the initiative for the Clairton agreement came from the international union. In 1971 the international pressed the local to make health and safety a local issue, thus giving the local the right to strike if the company failed to meet its demands. The local declined. In 1974 the international office, represented in large part by former Clairton president Dan Hannan, refused to approve a local settlement if it did not address the coke-oven issues. The agreement to negotiate was signed by I. W. Abel, the international president. Abel's special assistant, James Smith, played an integral role in the local negotiations. A number of the key issues were ultimately resolved by decision of the international office.

17
Many of these changes might have occurred eventually without the benefit of the Clairton agreement. On January 20, 1977, the Department of Labor officially adopted the OSHA Coke Oven Standard, which mandated many of the same engineering controls contained in the Clairton agreement. In effect, the agreement forced the company to implement these duplicate provisions a few years before it otherwise might have done so.

18
The Steelworkers' negotiators and the occupational hygienists that advised them recognized at the time of negotiation the risk they were running by increasing the crew size. They decided to accept the risk in the belief that it was not possible to clean up the coke ovens at all without increasing the size of the crew. (Personal communication, James English.)

19
The substantive standard applied in determining the validity of health and safety grievances is whether the employee is being "required to work under conditions which are unsafe or unhealthy beyond the normal hazard inherent in the operation in question." The employee is entitled to return to his job even if the arbitrator rules against him. If the arbitrator rules in his favor and the company has not provided alternative employment during the processing of the grievance (the company is not obligated to provide other employment), then the employee is entitled to back pay for the period in which he did not work or worked at a lower-rated job.

20
Hearings, Subcommittee on Labor, Committee on Labor and Public Welfare, U.S. Senate, *OSHA Review 1974*, Appendix II, Issue Papers, p. 1023. Unfortunately, later statistics are not available. OSHA has promulgated new guidelines in its field manual that are intended to speed

up agency response to employee complaints. The guidelines require investigation of "imminent danger" complaints within 24 hours, "serious" complaints within 3 working days, and all other complaints within 20 days. It is not clear what impact these guidelines have had on practice.

21
A health and safety grievance is any grievance filed under section 14 of the Basic Steel agreement. Section 14 is entitled "Safety and Health."

22
The Basic Steel Agreement specifies that "The Company and the Union will cooperate in the objective of eliminating accidents and health hazards. The Company shall make reasonable provisions for the safety and health of its employees at the plants during the working hours of employment. The Company, the Union and the employees recognize their obligation and/or rights under existing federal and state laws with respect to safety and health matters."

23
See for example D. Quinn Mills, *Industrial Relations and Manpower in Construction* (Cambridge, Mass.: MIT Press, 1972).

24
A 1976 Bureau of Labor Statistics survey of major collective bargaining agreements reported that fewer than 9 percent of bargaining agreements in construction referred to any kind of inspection. By comparison, over 50 percent of the manufacturing workforce is employed under agreements containing such provisions. ("Major Collective Bargaining Agreements: Safety and Health Provisions," U.S. Department of Labor bulletin 1425-16, p. 55).

25
See table 5.5 for an analysis of the health and safety provisions of 181 local collective bargaining agreements covering plumbers and pipefitters.

26
This statement does not suggest that training programs are necessarily effective in reducing the proportion of accidents caused by carelessness. The union has made no attempt to evaluate the impact of these programs on the accident rate.

27
All references in this section are to "Major Collective Bargaining Agreements: Safety and Health Provisions," Bureau of Labor Statistics bulletin 1425-16 (1976). This study examined 1,724 collective bargaining agreements, each covering 1,000 or more workers.

28
BLS bulletin, p. 12.

Notes to pp. 81–90

29
See T. Kochan, L. Dyer, and D. Lipsky, *The Effectiveness of Union-Management Safety and Health Committees* (Kalamazoo, Mich.: W. E. Upjohn Institute for Employment Research, 1977).

30
BLS bulletin, p. 50.

31
Provision negotiated by Babcock & Wilcox Co. Tubular Products Division, Beaver Falls, Pa., and United Steelworkers, August 1977.

32
If dangerous work is continually performed on a particular job there is not likely to be a hazard-pay provision, since the hazard-pay differential will be incorporated into the basic pay rate for the job classification.

33
Survey of Occupational Health Efforts of Fifteen Major Labor Unions (Washington, D.C.: Health Research Group, 1976). The unions surveyed included the Oil, Chemical and Atomic Workers; the United Mineworkers; the United Electrical Workers; the Communications Workers; the International Brotherhood of Painters; the United Auto Workers; the International Chemical Workers; the United Paperworkers; the United Steelworkers; the Textile Workers; the United Rubberworkers, AFGE, and AFSCME.

34
H. F. Howe, "Distribution of Occupational Physicians Among Industries," *Journal of Occupational Medicine* 11 (1969): 1190–1191.

Chapter 6

1
J. Dunlop, *Industrial Relations Systems* (New York: Holt, 1958).

2
It also might be possible to control coke-oven emissions by imposing a tax on management based on worker exposure to noxious emissions. An exposure tax, however, is nothing more than an implicit standard. A company would calculate the optimal level of emissions given the tax, and then would not exceed it.

3
The cost of strikes in steel has led to a number of innovative collective bargaining approaches to strike avoidance, such as the Experimental Negotiating Agreement instituted in 1973.

4
Not all health and safety bargaining is distributive. If labor forgoes other benefits to obtain improved working conditions, there has been a

bargained-for exchange and both sides are better off after the transaction. Similarly, management may benefit from negotiated improvements in health and safety conditions if its workmen's compensation premiums accurately reflect the changes in the risk to which the insured population is exposed. Because labor stands to gain at the bargaining table from refinement in the workmen's-compensation rating system, it is interesting that labor has not been more active in the movement for better experience rating.

5
Overall, the passage of the OSH Act has encouraged collective bargaining over health and safety issues because it has focused attention on occupational risks. Nonetheless, publication of a proposed OSHA regulation discourages both labor and management from negotiating an agreement dealing with the same hazard.

6
See Herrick, "Institutional Attitudes Toward Human Fulfillment Through Work," cited in N. Ashford, *Crisis in the Workplace: Occupational Disease and Injury* (Cambridge, Mass.: MIT Press, 1976), p. 91. A plurality of workers ranked health and safety first among topics demanding union attention.

7
Personal communication, James Fiore, former president of Clairton Local. The Clairton workers were not faced with precisely the same issue raised in the first question on the Kelman survey, which might explain part of the difference in response. Kelman asked workers whether they would forgo additional income for improved health conditions. At Clairton workers were faced with possible reductions in existing incomes. Perhaps more important, the Clairton workers differed in their view of the effectiveness of the proposed controls; they could not be sure that the controls would eliminate all danger from carcinogens.

8
In fact, such a local should be able to "afford" an even larger health and safety investment because of the differential in pre- and post-tax earnings.

9
See Kochan, Dyer, and Lipsky, p. 32.

10
Personal communication, J. Sheehan, Legislative Director of United Steelworkers.

11
See J. Kuhn, *Bargaining in Grievance Settlement* (New York: Columbia University Press, 1961), p. 34.

12
See L. Sayles and G. Strauss, *The Local Union* (New York: Harcourt Brace and World, 1967), p. 81.

Notes to pp. 96–100

13
George Haglund, unpublished Ph.D. thesis cited in Ashford, *Crisis in the Workplace*, p. 111. Haglund notes that the distribution of accidents is bimodal, suggesting that both age and experience contribute to the likelihood of an accident.

14
In the Kelman survey of OCAW workers cited earlier, 98 percent of local officers preferred elimination of the carcinogen to the $5,000 after-tax payment, as compared to only 89 percent of the rank and file surveyed. Furthermore, younger workers were more inclined to take the money than older workers. These findings are not necessarily inconsistent with the hypothesis expressed above in that the survey only measures worker response to identical exposure levels for the same risk. On an absolute basis, workers in high-risk positions may still be the most safety-conscious workers in the plant in relation to exposure level.

15
See Kuhn, *Bargaining in Grievance Settlement*.

16
A litigious worker recently sued the federal government, claiming that his injury was caused by OSHA's negligent failure to examine defective equipment. See *Blessing v. U.S.*, 447 F. Supp. 1160, Pennsylvania District Court, 1978.

17
See *Bryant v. United Mine Workers*, 467 F. 2d. 1, 6th Circuit, 1972; *Brough v. United Steelworkers of America*, 437 F. 2d. 748, 1st Circuit, 1971; *Helton v. Hake*, 386 F. Supp. 1027, Western District, Missouri, 1974. In *House v. Mine Safety Appliance Co.* (417 F. Supp. 939, Idaho District Court, 1976), the union was sued by the employer, who claimed that the union failed to discover hazards and was therefore liable to the employer for contribution in a damage action brought by an employee. To recover against his union, a worker must show that the union has breached its duty to fairly represent him. Fair representation only implies representation that is without discrimination, hostility, or arbitrariness. A mere showing of negligence will not sustain an action by a worker against his union.

18
Union officials would only admit to the existence of these agreements in private, for fear that knowledge of the agreements would encourage additional suits by workers.

19
Personal communication, Daniel Hannan, United Steelworkers.

20
Testimony of George Perkel, Research Director, Textile Workers Union, p. 606, and Frederick Mann, President, Local 502, United Auto Workers, p. 857, in U.S. Senate Occupational Safety and Health Act Hearings, Commit-

tee on Labor and Public Welfare, Subcommittee on Labor, 91st Congress, 1st and 2nd sessions on S. 2193 and S. 2788, 1970.

Chapter 7

1
Corn, "Status Report on OSHA," p. 4.

2
"OSHA Inspections," Programs and Policy Series, U.S. Department of Labor, Occupational Safety and Health Administration, report OSHA 2098, 1975, p. 2. Class training is supplemented with on-the-job training and individual study. With a few exceptions, OSHA inspectors are generalists who do not specialize by industry.

3
OSHA might consider treating a refusal by an employer to respond to a steward's request to comply with a regulation as evidence of a "willful violation of a standard" that exposes the employer to potentially larger fines.

4
Exemption is also desirable because it maintains the appearance of equity, represented by uniform standards, while exploiting some of the efficiency gained from letting labor and management strike their own health and safety bargains.

5
For example, a black worker injured because of a racially motivated failure on the part of the union to process his safety complaint could sue successfully on a "breach of duty of fair representation" theory. For a full discussion of the considerations in a "fair representation" suit see A. Cox, "Rights Under a Labor Agreement," *Harvard Law Review* 69 (1956): 601–657.

6
Personal communication, Ronald Coene, NIOSH Director of Office of Program Planning and Evaluation.

7
In the Clairton case, coke-oven emissions did not become a major contract issue until after publication of the steelworker mortality study in C. Redmond et al., "Long Term Mortality Study of Steelworkers, VI—Mortality From Malignant Neoplasms Among Coke Oven Workers," *Journal of Occupational Medicine* 14 (1972): 621–629.

8
Personal communication, Ronald Coene.

9
NIOSH has observed that union requests for health-hazard inspections increase just prior to negotiation of collective bargaining agreements (personal communication, Ronald Coene).

10
The pledge is most often tendered in return for an agreement from management to arbitrate disputes arising out of interpretation of the terms of the contract. Where management has agreed to arbitrate disputes and the union has not given an express no-strike pledge, the courts will imply a reciprocal no-strike pledge on the part of the union. See *Boys Markets Inc.* v. *Retail Clerks Local 770,* 398 U.S. 235, 1970.

11
For a more complete discussion of the right to refuse hazardous work see J. Atleson, "Threats to Health and Safety: Employee Self-Help Under the NLRA," *Minnesota Law Review* 59 (1975): 647–713 and N. Ashford and J. Katz, "Unsafe Working Conditions: Employee Rights Under the Labor Management Relations Act and the Occupational Safety and Health Act," *Notre Dame Lawyer* 52 (1977): 802–837. The legal analysis that follows draws heavily on these two articles.

12
29 U.S.C. sec. 157 and 29 U.S.C. sec. 158(a)(1), 1970.

13
Atleson, p. 660.

14
29 U.S.C. sec. 143, 1970.

15
Gateway Coal v. *United Mine Workers,* 414 U.S. 368, 387, 1973.

16
Ibid.

17
"We also disagree with the implicit assumption that the alternative to arbitration holds greater promise for the protection of employees. Relegating safety disputes to the arena of economic combat offers no greater assurance that the ultimate resolution will ensure employee safety. Indeed, the safety of the workshop would then depend on the relative economic strength of the parties rather than on an informed and impartial assessment of the facts." (414 U.S. 368, 379.)

18
See 29 CFR section 19.77.-12.

19
Whirlpool Corporation v. *Marshall,* 48 LW 4189, 1980.

20
48 LW 4194.

21
It can be argued that a worker who loses work time because an employer has not abated a life-threatening hazard should have the same claim to

back pay as a worker who loses work time because of actions by an employer that violate the terms of a collective bargaining agreement. Back pay is available in cases of racial discrimination as well as in cases of unfair labor practices.

22
In theory, industrial hygienists could also be trained as arbitrators. In practice this is likely to prove infeasible. Health and safety professionals would probably be viewed by management as biased in favor of workers. Moreover, as noted in chapter 1, many health and safety issues are inextricably intertwined with other bargaining issues.

Chapter 8

1
See Steven L. Yaffe, "Prohibitive Policy and the Implementation of the Endangered Species Act," Ph.D. diss., MIT, June 1979.

2
A. Nichols and R. Zeckhauser, "Government Comes to the Workplace: An Assessment of OSHA," *The Public Interest* 39 (1977): 39–69.

3
This idea was suggested to me by Michael O'Hare.

4
In delegating enforcement responsibility, it is of course important that the enforcing authority have the best interests of the protected constituency in mind. If it does not, then there is little to be gained from such a delegation. There is little reason, however, to believe that a federal regulatory agency will be more responsive to local interests than, say, a union or a local government. If lack of responsiveness is a legitimate concern, two alternatives might be considered. First, efforts can be made to stimulate responsiveness (e.g., giving individual workers the right to refuse hazardous work as a means of forcing union leadership to address health and safety issues). Second, the regulatory authority can set limits on the capacity of the enforcing authority to bargain away compliance. For example, absolute limits might be placed on the number of noise variances issued by a local government.

5
Regulations that have paternalistic dimensions (such as occupational safety and health standards, housing codes, and noise standards) often impose costs as well as benefits upon the protected constituency. For example, a noise standard that forced the closing of the only employer in town would certainly be a mixed blessing for the town's residents. In such cases, delegating enforcement authority permits the protected class to make these difficult choices for themselves.

Notes to pp. 127–130

6
See for example L. Susskind, J. Richardson, and K. Hildebrand, "Resolving Environmental Disputes; Approaches to Intervention, Negotiation, and Conflict Resolution," MIT Laboratory of Architecture and Planning report, June 1978.

7
For a more complete discussion of the uses of compensation in environmental decisionmaking, see M. O'Hare, " 'Not on My Block, You Don't'— Facility Siting and the Strategic Importance of Compensation," *Public Policy* 25 (1977): 407–458.

8
Unions can bind their members because of an act of Congress. The National Labor Relations Act provides that a properly elected union is the exclusive bargaining representative for workers in the bargaining unit; management is prohibited from negotiating with any other group.

Selective Bibliography

Ashford, N. A. *Crisis in the Workplace: Occupational Disease and Injury* (Cambridge, Mass.: MIT Press, 1976).

Ashford, N. A., et al. "Economic/Social Impact of Occupational Noise Regulations," testimony presented at OSHA hearings on the economic impact of occupational noise exposure, Washington, D.C. (September 1976).

Ashford, N. A., and Katz, J. "Unsafe Working Conditions: Employee Rights Under the Labor Management Relations Act and the Occupational Safety and Health Act," *Notre Dame Lawyer* 52 (1977): 802–837.

Atleson, J. "Threats to Health and Safety: Employee Self-Help Under the NLRA," *Minnesota Law Review* 59 (1975): 647–713.

Bacow, L. S. "Regulating Occupational Hazards Through Collective Bargaining," National Technical Information Service report ASPER/PUR-77/1914/A (Washington, D.C., 1978).

Baumol, W. J., and Oates, W. E. "The Use of Standards and Prices for Protection of the Environment," in P. Bohm and A. V. Kneese, eds., *The Economics of the Environment* (London: St. Martins, 1971), pp. 53–65.

Bok, D., and Dunlop, J. *Labor and the American Community* (New York: Simon and Schuster, 1970).

Bolt, Beranek and Newman, Inc. "Economic Impact Analysis of Proposed Noise Control Regulation," report prepared for U.S. Department of Labor Occupational Safety and Health Administration (1976).

Bureau of National Affairs. *The Job Safety and Health Act of 1970* (Washington, D.C., 1971).

———. *OSHA and the Unions: Bargaining for Job Safety and Health* (Washington, D.C., 1971).

Chamberlain, N. W. *Collective Bargaining* (New York: McGraw-Hill, 1951).

Chelius, J. "The Control of Industrial Accidents: Economic Theory and Empirical Evidence," *Law and Contemporary Problems* 38 (1974): 700–729.

———. "Expectations for OSHA's Performance: The Lessons of Theory and Empirical Evidence," Purdue University report (1975).

Corn, M. "Status Report on OSHA," U.S. Department of Labor (January 1977).

Cornell, N., Noll, R., and Weingast, B. "Safety Regulation," in H. Owen and C. E. Schutze, eds., *Setting National Priorities* (Washington, D.C.: Brookings Institution, 1976).

Cox, A. "Rights Under a Labor Agreement," *Harvard Law Review* 69 (1956): 601–657.

Di Pietro, A. "An Analysis of the OSHA Inspection Program in Manufacturing Industries 1972–1973," draft technical analysis paper, Office of the Assistant Secretary for Policy, Evaluation, and Research, U.S. Department of Labor (August 1976).

Selective Bibliography

———. "Data Needs for the Evaluation of OSHA's Net Impact," Office of Evaluation, Office of the Assistant Secretary for Policy, Evaluation, and Research, U.S. Department of Labor (August 1975).

Dunlop, J. *Industrial Relations Systems* (New York: Holt, 1958).

Foulkes, F. "Learning to Live with OSHA," *Harvard Business Review* (November–December 1973): 57–67.

Hale, A. R., and Hale, M. *A Review of the Industrial Accident Research Literature* (London: HMSO, 1972).

Health Research Group. *Survey of Occupational Health Efforts of Fifteen Major Labor Unions* (Washington, D.C., 1976).

Howe, H. F. "Distribution of Occupational Physicians Among Industries," *Journal of Occupational Medicine* 11 (1969): 1191.

Kalis, D. C. "Update on the Cost of OSHA Compliance," *Occupational Hazards* (August 1974): 42–44.

Kasper, D. M. "An Alternative to Workmen's Compensation," *Industrial and Labor Relations Review* 28 (1975): 535–548.

Kneese, A., and Schultze, C. *Pollution, Prices, and Public Policy* (Washington, D.C.: Brookings Institution, 1975).

Kochan, T., Dyer, L., and Lipsky, D. *The Effectiveness of Union-Management Safety and Health Committees* (Kalamazoo, Mich.: W. E. Upjohn Institute for Employment Research, 1977).

Kuhn, J. W. *Bargaining in Grievance Settlement* (New York: Columbia University Press, 1961).

Larson, A. *The Law of Workmen's Compensation*, 5 vols. (Albany, N.Y.: M. Bender, 1952–1976).

Mendeloff, J. "An Evaluation of the OSHA Program's Effect on Workplace Injury Rates: Evidence from California Through 1974," report prepared for the Office of the Assistant Secretary for Policy, Evaluation, and Research, U.S. Department of Labor, under contract B-9-M-5-2399 (July 1976).

Mendeloff, J. *Regulating Safety: An Economic and Political Analysis of Occupational Safety and Health Policy* (Cambridge, Mass.: MIT Press, 1979).

Miller, R. S. "The Occupational Safety and Health Act of 1970 and the Law of Torts," *Law and Contemporary Problems* 38 (1974): 612–640.

Mills, D. Q. *Industrial Relations and Manpower in Construction* (Cambridge, Mass.: MIT Press, 1972).

Morey, R. S. "The General Duty Clause of the Occupational Safety and Health Act of 1970," *Harvard Law Review* 86 (1973): 988–1005.

National Commission on State Workmen's Compensation Laws. *Compendium on Workmen's Compensation* (Washington, D.C., 1972).

———. *Report on the National Commission of State Workmen's Compensation Laws* (Washington, D.C., 1972).

Nichols, A., and Zeckhauser, R. "Government Comes to the Workplace: An Assessment of OSHA," *Public Interest* 49 (1977): 39–69.

O'Hare, M. " 'Not on My Block, You Don't'—Facility Siting and the Strategic Importance of Compensation," *Public Policy* 25 (1977): 407–458.

Oi, W. Y. "On the Economics of Industrial Safety," *Law and Contemporary Problems* 38 (1974): 669–699.

Oi, W. Y. "On Evaluating the Effectiveness of the OSHA Inspection Program," report submitted to U.S. Department of Labor under contract L-72-86 (May 1975).

Page, J. A., and O'Brien, M. *Bitter Wages* (New York: Grossman, 1973).

Peach, D. and Livernash, E. R. *Grievance Initiation and Resolution* (Cambridge, Mass.: Harvard University Press, 1974).

Pfeifer, C. M., et al. (eds.). "An Evaluation of Policy Related Research on Effectiveness of Alternative Methods to Reduce Occupational Illness and Accidents," report prepared by Behavioral/Safety Center, Columbia, Maryland (July 1974).

President's Report on Occupational Safety and Health. Annual reports for 1972, 1973, and 1974.

Prosser, W. L. *Handbook of the Law of Torts,* 4th edition (St. Paul, Minn.: West, 1971).

Rees, A. *The Economics of Trade Unions* (Chicago, Ill.: University of Chicago Press, 1962).

Russell, L. "Safety Incentives of Workmen's Compensation Insurance," *Journal of Human Resources* 8 (1974): 361–375.

Sands, P. E. "How Effective is Safety Legislation?" *Journal of Law and Economics* 11 (April 1968): 165–174.

Sayles, L. R., and Strauss, G. *The Local Union* (New York: Harcourt, Brace and World, 1967).

Schultze, C. *The Public Use of Private Interest* (Washington, D.C.: Brookings Institution, 1977).

Settle, R. "Benefits and Costs of the Federal Asbestos Standard," paper presented at Department of Labor Conference on Evaluating the Effects of the Occupational Safety and Health Program, Annapolis, Maryland, March 18–19, 1975.

Shafai-Sahrai, Y. "Determinants of Occupational Injury Experience," Division of Research, Graduate School of Business Administration, Michigan State University, East Lansing (1973).

Slichter, S., Healy, J., and Livernash, E. R. *The Impact of Collective Bargaining on Management* (Washington, D.C.: Brookings Institution, 1974).

Smith, R. S. "The Estimated Impact on Injuries of OSHA's Target Industry Program," Cornell University report (undated).

Selective Bibliography

———. "Evaluating the Impact of OSHA on Occupational Safety and Health," Office of the Assistant Secretary for Policy, Evaluation, and Research, U.S. Department of Labor (April 1975).

———. "The Feasibility of an 'Injury Tax' Approach to Occupational Safety," *Law and Contemporary Problems* 38 (1974): 730–744.

———. *The Occupational Safety and Health Act* (Washington, D.C.: American Enterprise Institute, 1976).

Stender, J. H. "Enforcing the Occupational Safety and Health Act of 1970: The Federal Government as a Catalyst," *Law and Contemporary Problems* 38 (1974): 641–650.

Surry, J. "Industrial Accident Research: A Human Engineering Appraisal," Labour Safety Council, Ontario Department of Labour, Toronto (1971).

Susskind, L., Richardson, J., and Hildebrand, K. "Resolving Environmental Disputes; Approaches to Intervention, Negotiation, and Conflict Resolution," Laboratory of Architecture and Planning, MIT (June 1978).

Thaler, R., and Rosen, S. "The Value of Saving a Life: Evidence from the Labor Market," paper presented at the NBER Conference on Income and Wealth, Household Production and Consumption, Washington, D.C., November 30, 1973.

U.S. Bureau of Labor Statistics. "Major Collective Bargaining Agreements in Selected Industries: Safety and Health Provisions Before and After the Occupational Safety and Health Act of 1970," report prepared by Office of Wages and Industrial Relations for Office of Occupational Health Statistics, Bureau of Labor Statistics (February 1975).

U.S. Congress, House of Representatives, Committee on Education and Labor, Select Subcommittee on Labor. Occupational Safety and Health Act of 1970 (Oversight and Proposed Amendments). Hearings, 93rd Congress, 2nd Session, Washington, D.C., 1975.

U.S. Congress, House of Representatives, Permanent Select Committee on Small Business, Subcommittee on Environmental Problems Affecting Small Business. The Effects of the Administration of the OSH Act on Small Business. Hearings, 93rd Congress, 2nd Session, Washington, D.C., 1974.

U.S. Congress, Senate, Committee on Labor and Public Welfare. *Legislative History of the Occupational Safety and Health Act of 1970* (Washington, D.C., 1971).

U.S. Congress, Senate, Committee on Labor and Public Welfare, Subcommittee on Labor. Occupational Safety and Health Act Review, 1974. Hearings, 93rd Congress, 2nd Session, Washington, D.C., 1974.

U.S. Department of Health, Education and Welfare. *Occupational Exposure to Coke Oven Emissions* (Washington, D.C.: NIOSH, 1973).

———. *Occupational Exposure to Inorganic Arsenic: New Criteria—1975* (Washington, D.C.: NIOSH, 1975).

U.S. Department of Labor. "Inflationary Impact Statement: Coke Oven Emissions," OSHA report, February 1976.

———. "Inflationary Impact Statement: Inorganic Arsenic," OSHA report (undated).

———. "Occupational Safety and Health Administration's Impact on Small Business," OSHA report (undated).

Viscusi, W. K. "Employment Hazards: An Investigation of Market Performance," Ph.D. diss., Harvard University, 1976.

———. "The Impact of Occupational Safety and Health Regulation," *Bell Journal of Economics* 10 (1979): 117–140.

"What It's Costing Industry to Comply with OSHA," *Occupational Hazards* (Fall 1974).

Wilson, J. Q. "The Politics of Regulation," in J. McKie, ed., *Social Responsibility and the Business Predicament* (Washington, D.C.: Brookings Institution, 1974).

Yaffe, S. L. "Prohibitive Policy and the Implementation of the Endangered Species Act." Ph.D. diss., MIT, 1979.

Zeckhauser, R. "Medical Insurance, A Case Study of the Tradeoffs Between Risk Spreading and Appropriate Incentives," *Journal of Economic Theory* (March 1970): 10–26.

Index

Abel, I. W., 68
Accidents
  causes, 37–38
  costs, 3
  incidence in U.S., 3, 24–25, 120
  reporting, 24, 27, 38–39
  seasonal factors, 4
  and training, 120
  and workers, 4, 33, 37–40
Administrative Procedures Act, 128
Air pollution, 9, 19–20
  control, 69–70
  standards, 29, 68–69, 70–71
Airline industry, 20
Allegheny County, Pa., 94
American National Standards Institute, 12
Arbitration, 99
  and compliance, 103
  costs, 118–119
  and health and safety complaints, 64, 70, 74–76, 87, 115–117
  training, 97, 117
Asbestos industry, 35–36
Asbestosis, 36
Automation, 4, 34
Automobile industry, 9–10, 61–67, 89

Bargaining
  attitude of parties, 90–93
  binding mechanisms, 129–131
  costs, 96–97
  and hazard abatement, 36, 51, 56–58, 61–87, 110–111
  implementation, 98–99, 116–117
  and information, 94
  and market mechanism, 89–90
  philosophical factors, 100–101, 111–112
  power relationships, 90
  privacy in, 129–130
  and standards, 18, 46, 68, 71
  and technology, 88–89, 97–99
  and workers, 94–96
"Bubble policy," 125
Building codes, 15, 19
Bureau of Labor Statistics, 27, 81
Byssinosis, 96

California, injury rates, 24–25, 40
Cancer, 3, 36, 94
Carcinogens, 33, 91, 92
  regulation, 124
Chemicals, 4, 15
Chrysler Corp., 9–10
Clairton agreement, 67–76, 86, 87, 90, 95, 98, 99
Coal mining, 7, 19
Coke-oven regulation, 67–76, 87, 88, 90, 94, 119, 127
  costs, 29, 30, 124
  worker attitudes, 91
"Command and control"
  strategies, 12–21
  and OSHA, 12–14, 79, 123
  and politics, 16–20
  and regulation, 14–16, 123–124
Compliance
  and competitive structure, 9–10
  costs, 30–32, 124
  incentives, 39–40, 47–49, 90, 104–106, 130
  and labor, 17, 90, 103
  and OSHA, 13, 45
  voluntary, 47
Construction industry, 5
  bargaining, 89–90
  hazard pay, 83–85
  labor, 77–81, 87
  regulation, 15, 16, 83, 87
Consumer-product safety, 6–7, 15
Consumer Product Safety Act, 15
Consumer Product Safety Commission, 8, 15
Contracting out, 33, 34
Copper mining, 7
Corn, Morton, 104
Costs
  accidents, 3
  bargaining, 96–97
  compliance, 30–32, 124
  and effectiveness, 43–47, 51, 124
  management, 29–32, 34, 41, 43, 47–49, 52, 92, 93
  measurement, 4
  mining regulation, 7
  monitoring, 53
  occupational disability, 3–4, 29–32

Index

Costs (continued)
  regulatory policies, 29–32, 34, 89
  research, 35
  and safety, 4, 35, 52, 92
  training, 80, 104–106
Cotton dust, 95, 96, 100
Courts, 13, 46, 115, 116
Criteria, 4
Criteria documents, 12, 40–41

Delaney Amendment, 124
Di Pietro, Aldonna, 26
Diversity
  of environments, 10–11, 19–20
  of hazards, 6–8
  of regulations, 8–10
  of workplaces, 5
Dunlop, John, 127

Efficiency, 18
  and standard setting, 43–47, 124, 131
Emissions, 9, 88
  and cancer, 94
  control, 69–70, 73–74
  standards, 18, 19, 29, 68–69, 70–71, 125
Employers. *See* Management
Endangered Species Act, 123–124
Energy conservation, 10, 15, 16, 19
Enforcement, 6, 13–14, 25, 38, 47
  and decentralization, 124–125
  discretion in, 108
  exemptions, 109–110
  and labor, 65, 102, 104–108
  legal proceedings, 126
Engineering, 33, 37
  and health and safety, 37, 69–70, 71, 73, 89, 99
Environmental Protection Agency, 15, 19, 125
Experts, 94, 97–99, 110

Fatigue, 4
Federal Register, 12, 13, 41
Fines, 13–14, 29, 39, 47–48, 49
  and workers, 109
First aid, 33, 83, 101
Florida, 10

Fluorocarbons, 9
Food and Drug Act, 124
Ford Motor Co., 9

Gases, exposure to, 82
Gateway Coal v. United Mine Workers, 115
General Motors Corp., 9, 61–67, 85, 86–87, 89
  and arbitration, 99, 101
  and inspection, 108
  negotiations, 89, 93
General Motors Institute, 61
Good Samaritan statutes, 109
Grievance procedures, 62–63, 70, 72, 74–75, 96–97, 99. *See also* Bargaining

Handicapped, 16
Harvard University, 85
Hazards, 32–35
  and NIOSH, 112–113
  and OSHA, 35–50
  research, 35
  and unions, 56–58, 61–87, 110–111, 113–114
  and wages, 32, 34, 36, 52, 78, 83–85, 86, 92
Health Hazard Evaluation Program, 112–113
Health Research Group, 85–86
Hearings, 13, 42–43, 45–46, 47, 126

Incentives
  and bargaining, 129, 131–132
  and management, 35, 36, 39–40, 52, 54
  and OSHA, 118
  and regulatory policy, 16, 18–20, 36, 39–40, 47–49
  and workers, 51, 69
Industrial Medical Association, 85–86
Information, 32, 36
  and bargaining, 94
  and workers, 53
Injuries. *See* Accidents
Inspections, 13
  costs, 101

Index

effectiveness, 47–49, 56, 79
and injury rate, 26, 27, 39–40
and OSHA, 13, 47–49, 56, 79
training, 17, 61–62, 64, 104–106
and unions, 61–62, 71–72, 89, 103–104
Insurance, 55, 92
Internal Revenue Code, 124
Iron ore mining, 7

Job rotation, 33, 46, 69
Journal of Occupational Medicine, 94

Kelman, Steven, 91–92

Labor
apprenticeship training, 79–81, 119–120
and health and safety regulation, 56–58, 60–87, 89–101, 104–106, 108, 110–112
inspections, 61–62, 71–72, 89, 103–108
leadership selection, 95–96
liability, 71–72, 97, 109–110
and management, 34, 76
and OSHA, 12–13, 56–58, 62, 65, 66, 80, 82–83, 86–87, 90, 97, 100, 103–108, 119–120, 124
and standard setting, 17, 41, 42, 45–46, 100
strikes, 113–114
and workmen's compensation, 55. See also Bargaining
Labor Management Relations Act, 114
Ladder safety, 12
Land reclamation, 7, 15
Lead smelting, 18, 89
Liability, 16
union, 71–72, 97, 109–110
Lung disease, 3, 36, 94

McGraw-Hill, 29
Maintenance, 48, 49, 70, 117
Management
costs to, 29–32, 34, 41, 43, 47–49, 92, 93
discretion, 93
incentives, 35
and labor, 34, 76
and medical facilities, 85
and OSHA, 12–13, 41, 45, 81
and safety representatives, 61, 63–65, 89, 101
and standard setting, 41, 42, 43, 45–46
and strikes, 113–114. See also Bargaining
Market mechanisms, 20
and bargaining, 89–90
and hazard abatement, 36, 52–53, 60
Medical facilities, 33, 83, 85–86
Mendeloff, John, 17, 24–25, 40
Mining industry, 7, 15, 36
Monitoring, 82–83, 94, 107
Morbidity and mortality studies, 3, 111–112

Nader, Ralph, 85
National Association of Manufacturers, 29–30, 47–48, 53
National Coal Policy Project, 127
National Contractors' Association, 80
National Institute for Occupational Safety and Health (NIOSH), 12, 40–41, 111–113
National Safety Council, 3
New York, N.Y., 39
Noise standards, 30, 45, 125
Nuclear industry, 3, 8
Nuclear Regulatory Commission, 8

Occupational Safety and Health Act, 8, 14, 16, 60
and research, 111–112
section 8(c), 3, 24
section 11(c), 116
section 20(6), 112–113
and steel industry, 68
Occupational Safety and Health Administration (OSHA)
costs, 29–32
critics, 51–53

OSHA (continued)
  effectiveness, 24–28, 30, 35–50, 75, 104
  enforcement powers, 13–14, 25, 39, 47, 65, 118
  inspections, 13, 47–49, 56, 79, 106–107
  and labor, 12–13, 56–58, 62, 65, 66, 80, 82–83, 86–87, 90, 97, 100, 103–109, 119–120, 124
  litigation, 46
  and management, 12–13, 41, 45, 81
  standards, 12–13, 14, 17–18, 40–47, 83
  training, 103, 106–107, 117
  and workers, 30
Office of Standards Development, 13, 42, 43
Office of Training and Education, 103
Oil, Chemical and Atomic Workers, 91–92

Pareto-optimal exchanges, 30, 44–45
Physical examinations, 46, 70, 83
Physicians. See Medical facilities
Polio, 5–6
Productivity
  and accidents and illness, 3–4
  and regulations, 9, 29, 30, 34
Protective devices, 33, 34, 83
  maintenance, 48, 49
Public Interest, The, 124
Punch press, 44, 45
Purdue University, 79–80

Radiation, 8
Regulatory policies
  costs, 29–32, 34, 89
  difficulties, 4–5, 7–8, 14
  effectiveness, 36–37, 43–47
  incentives, 16, 18–20, 36, 39–40, 47–49
  and labor, 101–102
  strategies, 14–16. See also "Command and control" strategies; Enforcement

Research, 35, 85, 111–112, 120
Respirator provisions, 82
Retail trade industries, 88

Salk vaccine, 5–6
Sanitation, 83
Schultze, Charles, 12
Sierra Club, 130
Smith, Robert, 26
Standards, 12–13, 14, 15–16, 17–18, 40–41
  and bargaining, 81–82
  complexity, 46
  costs, 24–28, 41
  disadvantages, 44–47
  uniformity, 50
Standards Advisory Committees, 12–13, 41–42, 43
State accident and health regulations, 16–17, 25, 39, 54–55, 83
Steel industry, 5, 67–76, 85
  arbitration in, 87, 115
  and regulation, 89
Stewards, union, 104–106
Strikes, 65, 113–114
Summer camps, 16
Sunshine laws, 128

Taft-Hartley Act, 114–115
Target Industry Program, 26, 28
Taxation
  accident, 16, 17, 51, 53–54
  and regulation, 18–19
Technology
  and bargaining, 88–89, 90–99
  and hazards, 36–37
  and industrial health, 4
Textile industry, 96, 100
Toxic substances, 53
Toxic Substances Control Act of 1975, 15, 123
Training
  apprenticeship, 79–81, 119–120
  arbitration, 99, 117
  costs, 80
  inspectors, 17, 61–62, 64, 104–106
  and OSHA, 103, 106–107, 117
  of union stewards, 104–106

Index

of workers, 33, 37, 61, 79–81, 86, 119–120
Trucking industry, 20

Unions. *See* Labor
United Association of Plumbers and Pipefitters, 61, 77–81
United Auto Workers, 61–67, 81, 85, 86, 87
  negotiations, 89, 94, 99, 101
United Rubber Workers, 85
U.S. Congress
  and OSHA, 13, 14, 51, 60, 123
  and regulation, 16, 18, 123
U.S. Department of Health, Education and Welfare, 15–16
U.S. Department of Labor, 12, 29, 115–116
U.S. Steel, 61, 68–76
U.S. Supreme Court, 13, 115, 116
United Steelworkers, 61, 67–76, 81, 85
  negotiations, 92, 94, 99
University of California, 98
University of North Carolina, 85
Uranium mining, 3
Utilities industry, 83

Ventilation standards, 82
Vinyl chloride, 95
Viscusi, W. Kip, 26–27, 29

Wages, 69
  and hazards, 32, 34, 36, 52, 78, 83–85, 86, 92
  and injury rate, 24–25
Walkoffs, 114–115
Walsh-Healey Public Contracts Act of 1936, 16–17
Water pollution, 10–11, 19, 125
Westwego, La., 3
Wilson, James Q., 18
Wisconsin, 39, 40
Work rules, 34, 57, 86
Workers
  accidents, 4, 33, 37–40
  age, 24, 27, 96
  and bargaining, 94
  fines, 109
  and hazard abatement, 52–53, 80–81, 113–116
  incentives, 51, 69
  minority, 27, 96
  mobility, 77–79
  priorities, 91–92, 95
  rights, 30, 90, 100, 110, 113–116
  security, 75–76, 83
  training, 33–37, 61, 79–81
  transfer, 46
Workmen's compensation, 16, 17, 25, 97
  and hazard abatement, 51, 54–55
  and labor, 55, 110
Workplaces
  diversity, 5
  exemption of, 106–108
  hazards, 36, 113–116
  inspection, 13, 32, 48, 56, 132
  standards, 16
  management responsibility, 76